T0228611

Mechanical Engineering: Design, Processes and Systems

Mechanical Engineering: Design, Processes and Systems

**Edited by
Rene Sava**

www.willfordpress.com

Published by Willford Press,
118-35 Queens Blvd., Suite 400,
Forest Hills, NY 11375, USA

Copyright © 2017 Willford Press

This book contains information obtained from authentic and highly regarded sources. Copyright for all individual chapters remain with the respective authors as indicated. All chapters are published with permission under the Creative Commons Attribution License or equivalent. A wide variety of references are listed. Permission and sources are indicated; for detailed attributions, please refer to the permissions page and list of contributors. Reasonable efforts have been made to publish reliable data and information, but the authors, editors and publisher cannot assume any responsibility for the validity of all materials or the consequences of their use.

Trademark Notice: Registered trademark of products or corporate names are used only for explanation and identification without intent to infringe.

ISBN: 978-1-68285-381-8

Cataloging-in-Publication Data

Mechanical engineering : design, processes and systems / edited by Rene Sava.
    p. cm.
Includes bibliographical references and index.
ISBN 978-1-68285-381-8
1. Mechanical engineering. 2. Machine design. 3. Machinery. I. Sava, Rene.
TJ145 .M43 2017
621--dc23

For information on all Willford Press publications
visit our website at www.willfordpress.com

Printed in the United States of America.

Contents

Permissions

List of Contributors

Index

Preface

Mechanical engineering is the study of mechanical systems, their design, manufacturing process and maintenance. Mechanical engineering has a variety of applications ranging from aerospace and aeronautics to mechatronics and nanotechnology. Research in this field strives to keep up with the latest technological innovations. This book is a compilation of research that seeks to advance the field of mechanical engineering. Different approaches, evaluations, methodologies and advanced studies on mechanical engineering have been included in this text. It brings forth some of the most innovative concepts and elucidates the unexplored aspects of this field. This book is a vital tool for all researching or studying mechanical engineering as it gives incredible insights into emerging trends and concepts. It attempts to assist those with a goal of delving into this field.

This book is a result of research of several months to collate the most relevant data in the field.

When I was approached with the idea of this book and the proposal to edit it, I was overwhelmed. It gave me an opportunity to reach out to all those who share a common interest with me in this field. I had 3 main parameters for editing this text:

1. Accuracy – The data and information provided in this book should be up-to-date and valuable to the readers.

2. Structure – The data must be presented in a structured format for easy understanding and better grasping of the readers.

3. Universal Approach – This book not only targets students but also experts and innovators in the field, thus my aim was to present topics which are of use to all.

Thus, it took me a couple of months to finish the editing of this book.

I would like to make a special mention of my publisher who considered me worthy of this opportunity and also supported me throughout the editing process. I would also like to thank the editing team at the back-end who extended their help whenever required.

Editor

A comparative numerical study of turbulence models for the simulation of fire incidents: Application in ventilated tunnel fires

Konstantinos G. Stokos[1]*, Socrates I. Vrahliotis[1], Theodora I. Pappou[2] and Sokrates Tsangaris[1]

*Corresponding author: Konstantinos G. Stokos, School of Mechanical Engineering, Section of Fluids, Laboratory of Biofluidmechanics & Biomedical Engineering, National Technical University of Athens, Heroon Polytechniou 9, Zografou, 15780 Athens, Greece
E-mail: kstokos@mail.ntua.gr

Reviewing editor: Duc Pham, University of Birmingham, UK

Abstract: The objective of this paper is to compare the overall performance of two turbulence models used for the simulation of fire scenarios in ventilated tunnels. Two Reynolds Averaged Navier–Stokes turbulence models were used; the low-Re k–ω SST and the standard k–ε model with wall functions treatment. Comparison was conducted on two different fire scenarios. The varied parameters were the heat release rate and the ventilation rate. Results predicted by the two turbulence models were also compared to the results produced from the commercial package Ansys Fluent. Quite faster simulations were performed using the k–ε turbulence model with wall functions and our findings, as to the basic characteristics of smoke movement, were in good agreement with Ansys Fluent ones.

Subjects: Fluid Dynamics; Computational Numerical Analysis; Heat Transfer; Computational Mechanics

Keywords: tunnel fire; ventilation; turbulence models comparison; numerical study

1. Introduction
Much attention has been paid on tunnel fires because of their negative consequences. A case of a fire accident in a tunnel constitutes an extremely dangerous situation for people who are inside it,

ABOUT THE AUTHORS
Our group's research and working areas relate to the development and application of software for computational mechanics. Specifically, the development and application for heat transfer problems (conduction, convection, and radiation), structural problems, aerodynamics, fluid–structure interaction, biofluid mechanics, numerical mesh generation, and optimization constitutes our field. Our team is involved in several transnational research projects in collaboration with organizations and companies. Many papers of our work have been published in scientific journals and international conferences.

PUBLIC INTEREST STATEMENT
Much attention has been paid on tunnel fires because of their negative consequences in human lives, repair cost, and cost due to the stop of tunnels operation. The methods used for the simulation of a tunnel fire are experimental, theoretical, and computational. Computational simulations cost less than experimental ones and they are more accurate compared to theoretical approaches.

Modeling of turbulence is a serious aspect of a CFD solver. The objective of this paper is the comparison of the overall performance of two turbulence models in two widely studied fire scenarios in a tunnel. Results predicted are also compared to the results of the general purpose commercial package Ansys Fluent. Both turbulence models produced accurate results and in good agreement to those of Ansys Fluent. Greater temperatures downstream from the heat source were calculated compared to experimental results, showing the importance of radiation and wall heat conduction modeling.

as hot toxic gases are produced. In addition to this, the repair cost and the cost due to the closing of tunnels operation are often huge.

The methods used for the simulation of a tunnel fire are experimental, theoretical, and computational. Full-scale experimental investigation is often prohibitive due to the necessary high cost, but it provides large amounts of reliable data. Many full-scale experiments for the Memorial tunnel are reported in the Memorial tunnel fire ventilation test program-test report (Massachusetts Highway Department, 1995). Small-scale experiments cost less and require careful choice of the scaling factors between the prototype and small scale (Lee & Ryou, 2005, 2006). Analytical methods and Computational Fluid Dynamics (CFD) simulations are the most affordable and permit the investigation of various alternative cases with proper modifications on the model. CFD is the most accurate between them, but a realistic CFD simulation of a fire scenario requires fluid dynamics, turbulence, radiation, wall conduction, and combustion numerical modeling.

Plenty CFD simulations of tunnel fires could be found in the literature, using commercial, open source and fewer of them research codes. When CFD simulation is applied, the selection of the appropriate turbulence model constitutes a serious aspect. Abanto, Reggio, Barrero, and Petro (2006) used Fluent and a research code to study smoke movement in a case of fire in an underwater tunnel. Turbulence was modeled with the standard $k-\varepsilon$ model. Hui, Li, and Lixin (2009) have modeled the longitudinal ventilation of the 4th Beijing subway line using CFX and have compared their results for the critical velocity to previous formulations proposed. They preferred the standard $k-\varepsilon$ model rather than LES because of computer load reasons. Lee and Ryou (2006) have studied the aspect ratio effect on smoke movement using Fire Dynamics Simulator (FDS) of the National Institute of Standards and Technology and compared their results to reduced-scale experimental ones. Turbulence was modeled using LES model. Hu, Peng, and Huo (2008) have studied the effect of the place of the fire in the critical velocity using FDS and LES turbulence model. Hu, Huo, Wang, and Yang (2007) have compared FDS with LES turbulence model results to full-scale experimental data with promising conclusions for the validity of FDS solver. Wu and Bakar (2000) applied an experimental and CFD investigation on the critical velocity formulations. For the CFD simulation, Fluent was used with the standard $k-\varepsilon$ turbulence model.

The most dangerous factor for human lives in tunnel fires is the excess of the concentration limits of the combustion products and not the extreme temperatures. Therefore, the appropriate ventilation system is required for the smoke control. One of the most important aims of a numerical model is the accurate prediction of the back-layering length and critical velocity produced by the current ventilation system. The distance of the smoke front from the heat source is the back-layering length. The critical velocity refers to longitudinal ventilation systems and is the lowest ventilation velocity that could prevent smoke back-layering. Longitudinal ventilation systems drive smoke to tunnel exit, ensuring the safe escape of passengers through the tunnel entrance and/or emergency exits. The ventilation system should prevent back-layering, but high ventilation velocities feed the fire with more oxygen, augmenting the heat release rate (Chow, 1998) and increase the resistance to the passengers, reducing the escaping rate (Hui et al., 2009). For the calculation of the critical velocity, mostly, Froude number-based analytical formulae have been used. Despite their simplicity, they do not account for some specific characteristics of each tunnel such as the existence of lateral evacuation hallways (Banjac & Nikolic, 2008), or the conditions of each fire scenario such as the place where accident happened or the obstructions that may exist. It was found that obstructions affect significantly the critical velocity (Kang, 2006; Oka & Atkinson, 1995). In case of not taking into account all these facts, huge investment costs for the ventilation system may be produced, or inadequate safety measures may be adopted. This fact makes CFD simulation the ideal method for the design of ventilation systems.

Our purpose is the comparison of the overall performance of the low-Re $k-\omega$ SST and the standard $k-\varepsilon$ with wall functions treatment turbulence models. The developed numerical solver is based on the incompressible Navier–Stokes and energy equations. Buoyancy force, due to temperature differences, was approximated by means of Boussinesq approximation. For both turbulence models, the necessary buoyancy source terms are included. Our results are also compared to those produced by the commercial package Ansys Fluent, using the standard k–ε turbulence model with wall functions under the same simplifications.

2. Mathematical model and numerical procedure

2.1. Test case and governing equations
The test case presented below is a widely studied one. Apte, Green, and Kent (1991) have carried out the experimental investigation. Fletcher, Kent, Apte, and Green (1994) have also presented experimental results and a numerical investigation using a steady-state approach and k–ε model with wall functions. Gao, Liu, Chow, and Fong (2004) have presented a numerical investigation using an unsteady approach and an LES turbulence model. Miloua, Azzi, and Wang (2011) also numerically studied this test case using FDS for the comparison of combustion models and wall boundary conditions, using LES turbulence model.

The tunnel geometry is described in Figure 1. A pool fire exists at 59.5 m from the entrance of the tunnel. The pool fire was assumed to be a cubic volumetric heat source, with the heat release rate being constant and having its maximum value from the beginning till the end of the simulation. The flow field in all cases tested was regarded as incompressible, because Mach number remained below 0.3. For the buoyancy forces, as a result of the temperature differences, Boussinesq approximation was adopted. Moreover, the viscous dissipation term in the energy equation was neglected, because the thermal energy due to viscous shear at low velocities is small (Gao et al., 2004). For the closure of the Reynolds Averaged Navier–Stokes (RANS) mean flow equations, two alternative turbulence models were used and compared; the low-Re k–ω SST and the less computationally demanding standard k–ε with wall functions treatment.

The general form of the conservation equations is:

$$\frac{\partial \phi}{\partial t} + \nabla \cdot \left(\bar{u}\phi \right) = \nabla \cdot \left(\Gamma_\phi \nabla \phi \right) + S_\phi \tag{1}$$

The conserved quantity "ϕ", the diffusion coefficient "Γ_ϕ" and the source term "S_ϕ" for each one of the conservation equations are given in Table 1 in tensorial notations.

The flow variables are the pressure "p," the Cartesian velocity components "u," "v," and "w" in the "x," "y," and "z" directions, respectively, and the temperature "T." Turbulence modeling inserts two additional dependent variables. The kinetic energy of turbulence "k" and turbulence dissipation "ε" when standard k–ε model is used and the turbulence kinetic energy "k" and specific dissipation rate "ω" when k–ω SST is used. The term "$g\beta_T \Delta T$" in the "z" direction momentum equation is the buoyancy force term, where "g" is the gravitational acceleration, "β_T" is the thermal expansion coefficient, and "ΔT" is the subtraction of the ambient temperature T_o (which is equal to 26°C) from

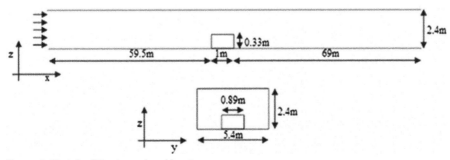

Figure 1. Sketch of the tunnel and heat source.

the local temperature. "ρ" is the fluid density at the ambient temperature and "C_p" is the specific heat at constant pressure. "$\dot{q}_c$" is the heat release rate per unit volume. Effective kinematic viscosity "ν_{eff}" is the sum of the molecular kinematic viscosity "ν" and the turbulence kinematic viscosity "ν_t". Turbulence kinematic viscosity is calculated by $\nu_t = c_\mu k^2/\varepsilon$ when standard $k-\varepsilon$ is used and $\nu_t = \frac{\alpha_1 k}{\max(\alpha_1 \omega, \Omega F_2)}$ when $k-\omega$ SST is used. For more information on the $k-\omega$ SST turbulence model readers are referred to (Menter, 1993). "Pr" is the Prandtl number and "Pr_t" is the turbulence Prandtl number for temperature. "σ_k", "σ_ε", and "σ_ω" are turbulence model constants (Barakos, Mitsoulis, & Assimacopoulos, 1994; Menter, 1993). Terms "P_k" and "G_k" correspond to shear and buoyancy production rates of the turbulence kinetic energy respectively. The inclusion of the "G_k" term in the turbulence conservative equations is of vital importance, since without this extremely false back-layering lengths may be calculated (Fletcher et al., 1994). "CD_ω" is the cross-diffusion term (Menter, 1993).

2.2. Boundary and initial conditions

Two different test cases were simulated, varying in the total heat release rate of the heat source and the ventilation rate. The total heat release rate and the ventilation velocity for these two cases are given in Table 2.

At the inlet of the tunnel uniform velocity profile was prescribed, with the "u" velocity being equal to the ventilation velocity while "v" and "w" velocities were equal to zero ($v = w = 0$). At the outlet of the tunnel, pressure was prescribed and set equal to the ambient pressure. At the tunnel walls

Table 1. Coefficients of Equation 1 and constants values

Conservation equation	ϕ	Γ_ϕ	S_ϕ
Continuity	1	0	0
x-momentum	u	ν_{eff}	$-\frac{1}{\rho}\frac{\partial p}{\partial x} + \frac{\partial}{\partial x_j}\left(\nu_{eff}\frac{\partial u_j}{\partial x}\right)$
y-momentum	v	ν_{eff}	$-\frac{1}{\rho}\frac{\partial p}{\partial y} + \frac{\partial}{\partial x_j}\left(\nu_{eff}\frac{\partial u_j}{\partial y}\right)$
z-momentum	w	ν_{eff}	$-\frac{1}{\rho}\frac{\partial p}{\partial z} + \frac{\partial}{\partial x_j}\left(\nu_{eff}\frac{\partial u_j}{\partial z}\right) + g\beta_T \delta T$
Energy	T	$\frac{\nu}{Pr} + \frac{\nu_t}{Pr_t}$	$\frac{\dot{q}_c}{\rho C_p}$
k-equation[a]	k	$\nu + \frac{\nu_t}{\sigma_k}$	$P_k + G_k - \varepsilon$
ε-equation	ε	$\nu + \frac{\nu_t}{\sigma_\varepsilon}$	$[C_{\varepsilon1}(P_k + C_{\varepsilon3}G_k) - C_{\varepsilon2}\varepsilon]\frac{\varepsilon}{k}$
k-equation[b]	k	$\nu + \frac{\nu_t}{\sigma_k}$	$P_k + G_k - \beta^*\omega k$
ω-equation	ω	$\nu + \frac{\nu_t}{\sigma_\omega}$	$\gamma\frac{\omega}{k}P_k + CD_\omega + \gamma C_{\varepsilon3}\frac{\omega}{k}G_k - \beta\omega^2$

$\nu_{eff} = \nu + \nu_t, P_k = \nu_t\left(\frac{\partial u_i}{\partial x_j} + \frac{\partial u_j}{\partial x_i}\right)\frac{\partial u_i}{\partial x_j},\quad G_k = -\frac{\nu_t}{Pr_t}g\beta_T\frac{\partial T}{\partial z},\quad CD_\omega = 2(1-f_1)\sigma_{\omega2}\frac{1}{\omega}\frac{\partial k}{\partial x_j}\frac{\partial \omega}{\partial x_j}$

$\rho = 1.18\frac{kg}{m^3}, g = 9.81\frac{m}{s^2}, \beta_t = 3.34\cdot10^{-3}\frac{1}{k}, cp = 1{,}006\frac{J}{kgK}, Pr = 0.71, Pr_t = 0.9, C_{\varepsilon1} = 1.44, C_{\varepsilon2} = 1.92, C_{\varepsilon3} = 1.0$

[a]Standard $k-\varepsilon$ turbulence model is used.

[b]$k-\omega$ SST turbulence model is used.

Table 2. Cases data and calculated results

Case	Ventilation velocity (m/s)	Total heat release rate (MW)	Numerical approach	Flame angle (°)	Back-layering length
1	$U_1 = 0.85$	$\dot{Q}_1 = 2.57$	Present solver	8	Till entrance
			Ansys Fluent	9	Till entrance
2	$U_2 = 2$	$\dot{Q}_2 = 2.29$	Present solver	58	3.2 m
			Ansys Fluent	58	3.7 m

no-slip boundary conditions were applied and they were assumed to be adiabatic, which corresponds to the worst situation. When low-Re k–ω SST turbulence model was applied, turbulence variables at the walls were

$$k=0, \quad \omega=10\frac{6\nu}{\beta_1(\Delta y)^2} \tag{2}$$

where "Δy" is the distance to the next point away from the wall. When high-Re k–ε turbulence model was applied, appropriate wall functions were used leading to coarser numerical meshes and less time demanding calculations. The values for "u", "v", "w", "T", "k", and "ε" were calculated using the following wall functions.

$$u^+ = \begin{cases} \frac{1}{\kappa}\ln y^+ + 5.5, & y^+ \geq 11.6 \\ y^+, & y^+ \leq 11.6 \end{cases} \tag{3}$$

$$T^+ = \Pr y^+ e^{-\Gamma} + \left[2.12\ln\left(1+y^+\right)+\beta(\Pr)\right]e^{-1/\Gamma} \tag{4a}$$

$$\beta(\Pr) = \left(3.85\,\Pr^{1/3}-1.3\right)^2 + 2.12\ln(\Pr) \quad \Gamma = \frac{0.01\left(\Pr y^+\right)^4}{1+5\,\Pr^3 y^+} \tag{4b}$$

$$y^+ = \frac{y u_\tau}{\nu} \tag{5}$$

$$u^+ = \frac{u}{u_\tau} \tag{6}$$

$$T^+ = \frac{T_w - T}{T_\tau} \tag{7}$$

$$u_\tau = \sqrt{\rho \tau_w} \tag{8}$$

$$T_\tau = \frac{q_w}{\rho C_p u_\tau} \tag{9}$$

$$k = \frac{u_\tau^2}{\sqrt{C_\mu}} \tag{10}$$

$$\varepsilon = \frac{u_\tau^3}{\kappa y} \tag{11}$$

For $y^+ \leq 11.6$ the first grid point is in the viscous sub-layer and for $y^+ > 11.6$ in the logarithmic region. It is desirable that the first grid point belongs to the logarithmic region of the boundary layer, where the viscous effects are weaker compared to the turbulent ones (Launder & Spalding, 1974). For the temperature near the wall, the formulation proposed by Kader was used (Kader, 1981), which is valid in all zones of the boundary layer and for a large range of Prandtl numbers.

As initial condition, the converged to steady-state flow field of the isothermal case (without the heat source) was used. The temperature was equal to the ambient temperature.

2.3. Numerical methodology
The solution procedure of the mean flow equations has been analyzed in details by Stokos, Vrahliotis, Pappou, and Tsangaris (2012, 2013) and Vrahliotis, Pappou, and Tsangaris (2012). A node centered finite volume discretization technique has been developed for hybrid numerical meshes. The solu-

tion of the incompressible flow equations was based on the artificial compressibility method. A Roe's approximate Riemann solver was used for the evaluation of the convective fluxes. For the discretization of the viscous fluxes a central scheme was used. For the evaluation of the convective fluxes of the turbulence equations a first-order upwind scheme was applied, as proposed by Menter (1993). For the calculation of turbulence equations viscous fluxes, the same method used for the mean flow equations was applied. For the temporal discretization, a dual-time stepping approach was adopted. Specifically, an implicit second-order backward difference scheme was implemented for the physical-time marching and an implicit first-order scheme for the pseudo-time marching. To pass from one physical time step to another convergence in pseudo-time needed to be achieved. The mean flow and turbulence equations were loosely coupled. The mean flow equations were first solved with the turbulence kinematic viscosity fixed. Then, the turbulence equations were solved with the velocity and temperature fields fixed. The resulting non-linear algebraic system of equations was linearized using Newton's method. The arising linear system was solved using the Jacobi iterative method with suitable under-relaxation. Parallel processing was used for the acceleration of the computations.

2.4. Numerical mesh

The computing domain was composed of 481,950 nodes and 449,000 hexahedrons when standard $k-\varepsilon$ turbulence model with wall functions was used (Figure 2). The first layer thickness was approximately equal to 0.01. When low-Re $k-\omega$ SST was used the computing mesh was composed of 1,094,252 nodes and 1,075,400 hexahedrons (Figure 3). The first layer thickness was of the order of 10^{-5} so that the $y^+ < 2$ condition is fulfilled and the growing factor was equal to 1.2. Both numerical meshes were denser near the heat source. The physical time step was set equal to 0.01 s.

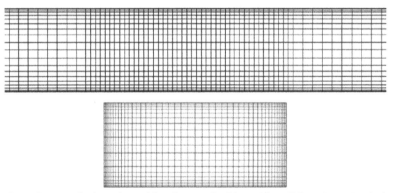

Figure 2. Numerical mesh used for the standard $k-\varepsilon$ with wall functions simulations.

Notes: Section $y = 2.7$ m near the vicinity of the heat source (top figure). Cross-section $x = 60$ m (bottom figure).

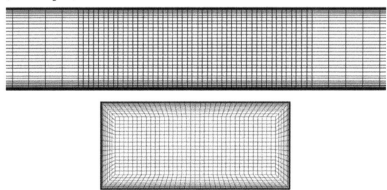

Figure 3. Numerical mesh used for the low-Re $k-\omega$ SST simulations.

Notes: Section $y = 2.7$ m near the vicinity of the heat source (top figure). Cross-section $x = 60$ m (bottom figure).

3. Results and discussion

3.1. Transient results

When no concentrations of smoke are calculated and radiative heat transfer is not considered, it could be assumed that smoke movement is analogous to the temperature field. In Figures 4 and 5, the evolution of the flame and temperature field for cases 1 and 2, respectively, for the first 10 s are presented. It is obvious that the predicted smoke movement through time by the two turbulence models is similar.

In a tunnel fire case surrounding medium (air and smoke) in the vicinity of the heat source is heated and raises up till the ceiling of the tunnel. Then having reached the ceiling it moves to the side walls and along the ceiling to the tunnel exit and the tunnel inlet forming the back-layering length. Reaching the side walls, smoke moves downward to the ground. In Figures 6 and 7 velocity vectors predicted by both turbulence models are given at characteristic sections and moments. Velocity vectors reveal smoke movement. It is obvious that smoke requires less than 1 s to reach the ceiling and less than 5 s to reach the side walls for both cases. Similar flow patterns were predicted by the turbulence models.

In Figure 8, we present the predicted using both turbulence models temperature vertical profiles at 18 and 30 m downstream from the heat source for case 2, 20 s after fire breaking. Small differences (less than 6%) are observed for the vertical temperature profiles.

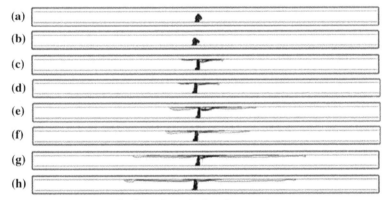

Figure 4. Temperature isolines through time for case 1.

Notes: (a) Standard $k-\varepsilon$ at 0.5 s; (b) Low-Re $k-\omega$ SST at 0.5 s; (c) Standard $k-\varepsilon$ at 2 s; (d) Low-Re $k-\omega$ SST at 2 s; (e) Standard $k-\varepsilon$ at 5 s; (f) Low-Re $k-\omega$ SST at 5 s; (g) Standard $k-\varepsilon$ at 10 s; and (h) Low-Re $k-\omega$ SST at 10 s.

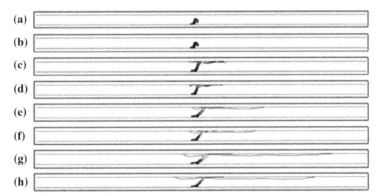

Figure 5. Temperature isolines through time for case 2.

Notes: (a) Standard $k-\varepsilon$ at 0.5 s; (b) Low-Re $k-\omega$ SST at 0.5 s; (c) Standard $k-\varepsilon$ at 2 s; (d) Low-Re $k-\omega$ SST at 2 s; (e) Standard $k-\varepsilon$ at 5 s; (f) Low-Re $k-\omega$ SST at 5 s; (g) Standard $k-\varepsilon$ at 10 s; and (h) Low-Re $k-\omega$ SST at 10 s.

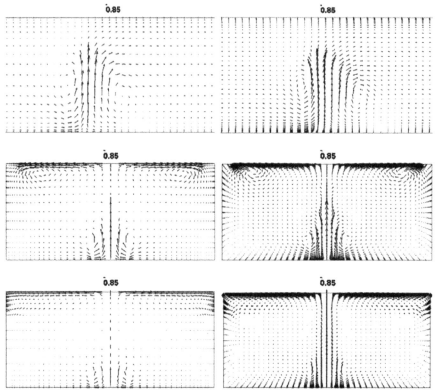

Figure 6. Velocity vectors at characteristic sections and moments for case 1 predicted by the standard k–ε model (left column) and the low-Re k–ω SST model (right column).

Notes: Section y = 2.7 m near the heat source at 0.5 s (top row), cross-section x = 60 m at 1 s (middle row), and cross-section x = 60 m at 5 s (bottom row).

In Figure 9, we present for case 2 the convergence history for the temperature corrections in pseudo-time for three physical time steps, using the standard k–ε turbulence model. The starting point is t = 2 s. It is obvious that less than 250 pseudo-time steps are required for each physical time step. However, it should be noted that the first 2 s are the most numerically crucial and more pseudo-time steps are required for pseudo-time convergence compared to the rest of the simulation. Same smooth convergence occurred and for the other dependent variables.

3.2. Steady state results

Gao et al. (2004) claim that flame shape is defined by the maximum temperature gradients. In the literature, many definitions for the flame angle have been found (Anderson et al., 2006). In Figures 10 and 11, flame shapes and the definition used for the calculation of the flame angle are given. Flame angle β_t is defined by the vertical line passing through the core of the heat source and the line connecting the core of the heat source with the upper point of the flame. In the same figures, temperature fields in the vicinity of the heat source are compared to those predicted by Ansys Fluent. The greatest temperature values, flame shapes, flame angles and back-layering lengths seem to agree satisfactorily. Flame angles and back-layering lengths predicted by our solver and Ansys Fluent are given in Table 2. It is evident that the backflow is less and flame tilt greater for higher ventilation velocities. The time needed for the smoke front to reach the steady-state back-layering length was about 50 and 15 s for case 1 and case 2, respectively.

Radiation plays a significant role in a case of fire in a tunnel. The fraction of the heat release from a heat source in a tunnel fire in form of radiation is in the order of 20–50% (Bettelini, 2001; Hostikka, 2008; Grant, Jagger, & Lea, 1998). However, the solution of the radiative heat transfer equation is time consuming because a great number of radiation intensities have to be calculated at each computational node. More than 100 solid angles are utilized for each node at a finite volume-based

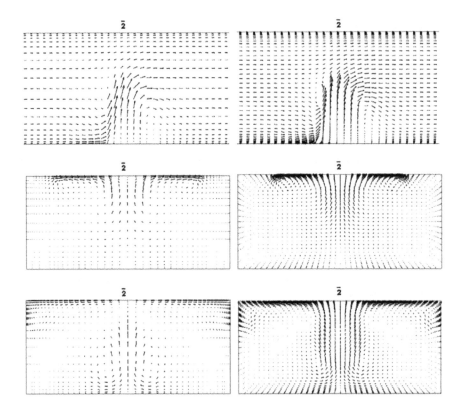

Figure 7. Velocity vectors at characteristic sections and moments for case 2 predicted by the standard k–ε model (left column) and the low-Re k–ω SST model (right column).

Notes: Section y = 2.7 m near the heat source at 0.5 s (top row), cross-section x = 62 m at 1 s (middle row) and cross-section x = 62 m at 5 s (bottom row).

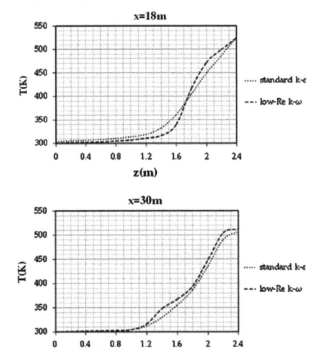

Figure 8. Case 2, 20 s after fire breaking.

Notes: Vertical temperature profiles at a distance of 18 m (top figure) and 30 m (bottom figure) downstream from the heat source.

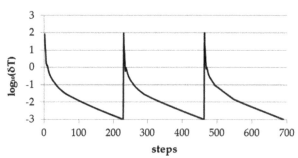

Figure 9. Convergence curve.

Notes: Temperature corrections in pseudo-time for three physical time steps with starting point $t = 2$ s.

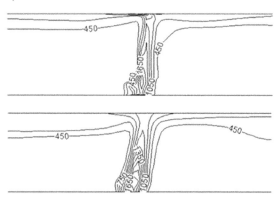

Figure 10. Flame shape comparison for case 1.

Notes: Present solver with the $k-\omega$ model (top figure), Ansys Fluent with the $k-\varepsilon$ model (bottom figure).

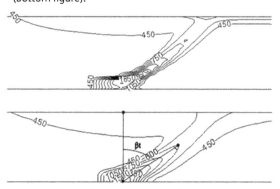

Figure 11. Flame shape comparison for case 2.

Notes: Present solver with the $k-\varepsilon$ model (top figure), Ansys Fluent with $k-\varepsilon$ model (bottom figure). Definition of flame angle.

method. Consequently, heat transfer due to radiation is often not taken into account, or the assumed amount of radiation loss is subtracted from the heat release rate (Kang, 2006). In Figure 12, temperature profiles are given 18 and 40 m downstream of the heat source. Greater values for the temperature are computed compared to the experimental values of Fletcher et al. (1994). This discrepancy is attributed to the omission of radiation modeling, according to the aforementioned role of radiation, and the omission of heat conduction inside tunnel wall. Similar discrepancies were calculated by Gao et al. (2004) who applied the same simplifications. However, the curves calculated are of the same "s" form. Temperature increases with increasing height and reaches its maximum values inside the smoke plume. Both turbulence models calculated approximate temperature profiles 18 and 40 m downstream from the heat source. Experimental investigation also showed approximate temperature profiles 18 and 40 m from the heat source.

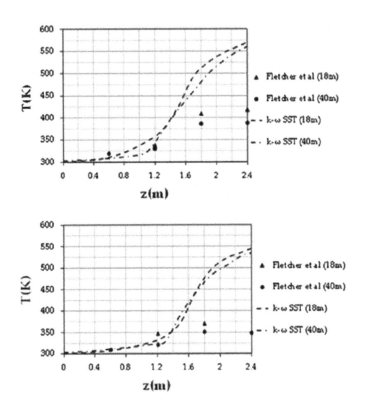

Figure 12. Comparison of temperature profiles along height when steady state was reached.

Notes: Case 1 (top figure) and case 2 (bottom figure).

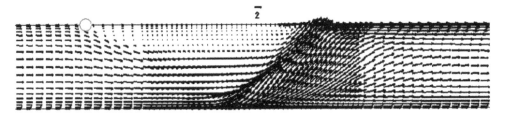

Figure 13. Velocity vectors at section y = 2.7 m and stagnation point (red circle).

Velocity vectors for case 2 with the standard $k-\varepsilon$ model when steady state was achieved are presented in Figure 13. The stagnation point which defines the back-layering length is marked with a red circle.

Finally, as far as computation time is concerned, a speedup approximately equal to 2.5, at each pseudo-time step, was gained when $k-\varepsilon$ turbulence model with wall functions was used compared to the low-Re $k-\omega$ SST turbulence model. This fact is mostly a result of the less dense grid required from the $k-\varepsilon$ turbulence model due to the use of wall functions.

4. Conclusions

The performance of the standard $k-\varepsilon$ with wall functions and the low-Re $k-\omega$ SST turbulence models was compared in terms of accuracy, stability, and calculation time. Flow pattern and temperature fields were in good agreement. Our results were also compared to the results produced by the commercial general purpose code Ansys Fluent. Back-layering lengths and flame angles for both cases agreed satisfactorily with Ansys Fluent ones. However, the omission of radiation and wall conduction modeling led to higher temperatures compared to experimental results. Simulations including both of these models will be presented in a future article with improved calculated temperature fields.

Acknowledgments
Simulations were performed on the high-performance platform "VELOS" of the Laboratory of Thermal Turbomachines, Parallel CFD and Optimization Unit, School of Mechanical Engineering of the National Technical University of Athens.

Funding
This work was funded under the EUROSTARS Project (E!5292) "Structural and Aerodynamic Design of TUNnels under Fire Emergency Conditions" from: the Greek General Secretariat for Research and Technology, Bundesministerium fur Bildung and Forschung (BMBF), FiDES DV-Partner GmbH, Germany and ELXIS Engineering Consultants SA. It was also funded from the Greek State Scholarships Foundation.

Author details
Konstantinos G. Stokos[1]
E-mail: kstokos@mail.ntua.gr
Socrates I. Vrahliotis[1]
E-mail: svrah@mail.ntua.gr
Theodora I. Pappou[2]
E-mail: dora@fides-dvp.de
Sokrates Tsangaris[1]
E-mail: sgt@fluid.mech.ntua.gr

[1] School of Mechanical Engineering, Section of Fluids, Laboratory of Biofluidmechanics & Biomedical Engineering, National Technical University of Athens, Heroon Polytechniou 9, Zografou, 15780 Athens, Greece.

[2] FIDES DV-Partner Beratungs-und Vertiebs-GmbH, Munich, Germany.

References

Abanto, J., Reggio, M., Barrero, D., & Petro, E. (2006). Prediction of fire and smoke propagation in an underwater tunnel. *Tunnelling and Underground Space Technology, 22*, 90–95.

Anderson, W., Pastor, E., Butler, B., Catchpole, E., Dupuy, J. L., Fernandes, P., ... Ventura, J. (2006, November 27–30). Evaluating models to estimate flame characteristics for free-burning fires using laboratory and field data. In *5th International Conference on Forest Fire Research*. Coimbra, Portugal.

Apte, V. B., Green, A. R., & Kent, J. H. (1991). Pool fire plume flow in a large-scale wind tunnel. *Fire Safety Science, 3*, 425–434. http://dx.doi.org/10.3801/IAFSS.FSS.3-425

Banjac, M., & Nikolic, B. (2008). Numerical study of smoke flow control in tunnel fires using ventilation systems. *FME Transactions, 36*, 145–150.

Barakos, G., Mitsoulis, E., & Assimacopoulos, D. (1994). Natural convection flow in a square cavity revisited: Laminar and turbulent models with wall functions. *International Journal for Numerical Methods in Fluids, 18*, 695–719. http://dx.doi.org/10.1002/(ISSN)1097-0363

Bettelini, M. (2001, September 17–18). CFD for tunnel safety. In Fluent users' meeting. Bingen.

Chow, W. K. (1998). On smoke control for tunnels by longitudinal ventilation. *Tunnelling and Underground Space Technology, 13*, 271–275. http://dx.doi.org/10.1016/S0886-7798(98)00061-3

Fletcher, D. F., Kent, J. H., Apte, V. B., & Green, A. R. (1994). Numerical simulations of smoke movement from a pool fire in a ventilated tunnel. *Fire Safety Journal, 23*, 305–325. http://dx.doi.org/10.1016/0379-7112(94)90033-7

Gao, P. Z., Liu, S. L., Chow, W. K., & Fong, N. K. (2004). Large eddy simulations for studying tunnel smoke ventilation. *Tunnelling and Underground Space Technology, 19*, 577–586. http://dx.doi.org/10.1016/j.tust.2004.01.005

Grant, G. B., Jagger, S. F., & Lea, C. J. (1998). Fires in tunnels. *Philosophical Transactions: Mathematical, Physical and Engineering Sciences, 356*, 2873–2906.

Hostikka, S. (2008). *Development of fire simulation models for radiative heat transfer and probabilistic risk assessment* (PhD Thesis). VTT Technical Research Centre of Finland, Espoo. ISBN: 978-951-38-7099-7.

Hu, L. H., Huo, R., Wang, H. B., & Yang, R. X. (2007). Experimental and numerical studies on longitudinal smoke temperature distribution upstream and downstream from the fire in a road tunnel. *Journal of Fire Sciences, 25*, 23–43. http://dx.doi.org/10.1177/0734904107062357

Hu, L. H., Peng, W., & Huo, R. (2008). Critical wind velocity for arresting upwind gas and smoke dispersion induced by near-wall fire in a road tunnel. *Journal of Hazardous Materials, 150*, 68–75. http://dx.doi.org/10.1016/j.jhazmat.2007.04.094

Hui, Y., Li, J., & Lixin, Y. (2009). Numerical analysis of tunnel thermal plume control using longitudinal ventilation. *Fire Safety Journal, 44*, 1067–1077. http://dx.doi.org/10.1016/j.firesaf.2009.07.006

Kader, B. A. (1981). Temperature and concentration profiles in fully turbulent boundary layers. *International Journal of Heat and Mass Transfer, 24*, 1541–1544. http://dx.doi.org/10.1016/0017-9310(81)90220-9

Kang, K. (2006). Computational study of longitudinal ventilation control during an enclosure fire within a tunnel. *Journal of Fire Protection Engineering, 16*, 159–181. http://dx.doi.org/10.1177/1042391506056737

Launder, B. E., & Spalding, D. B. (1974). The numerical computation of turbulent flows. *Computer Methods in Applied Mechanics and Engineering, 3*, 269–289. http://dx.doi.org/10.1016/0045-7825(74)90029-2

Lee, S. R., & Ryou, H. S. (2005). An experimental study of the effect of the aspect ratio on the critical velocity in longitudinal ventilation tunnel fires. *Journal of Fire Sciences, 23*, 119–138. http://dx.doi.org/10.1177/0734904105044630

Lee, S. R., & Ryou, H. S. (2006). A numerical study on smoke movement in longitudinal ventilation tunnel fires for different aspect ratio. *Building and Environment, 41*, 719–725. http://dx.doi.org/10.1016/j.buildenv.2005.03.010

Massachusetts Highway Department. (1995). *Memorial tunnel fire ventilation test program comprehensive test report*.

Menter, F. R. (1993, July 6–9). Zonal two equation k-ω turbulence models for aerodynamic flows. In *AIAA 24th Fluid Dynamics Conference*. Orlando, FL, AIAA 93-2906.

Miloua, H., Azzi, A., & Wang, H. Y. (2011). Evaluation of different numerical approaches for a ventilated tunnel fire. *Journal of Fire Sciences, 29*, 403–429. http://dx.doi.org/10.1177/0734904111400976

Oka, Y., & Atkinson, G. T. (1995). Control of smoke flow in tunnel fires. *Fire Safety Journal, 25*, 305–322. http://dx.doi.org/10.1016/0379-7112(96)00007-0

Stokos, K. G., Vrahliotis, S. I., Pappou, Th. I., & Tsangaris, S. (2012, July 4–7). Development and validation of a 3-D Navier–Stokes solver including heat transfer and natural convection. In *Proceedings of 5th International Conference*

from *Scientific Computing to Computational Engineering,
5th IC-SCCE*. Athens Greece.

Stokos, K. G., Vrahliotis, S. I., Pappou, Th. I., & Tsangaris, S.
(2013, May 25–27). Development and validation of a
Navier–Stokes solver including heat transfer and mixed
convection. In *Proceedings of 10th International Congress
on Mechanics, 10th HSTAM*. Chania, Crete, Greece.

Vrahliotis, S., Pappou, Th, & Tsangaris, S. (2012). Artificial
compressibility 3-D Navier–Stokes solver for unsteady

incompressible flows with hybrid grids. *Engineering
Applications of Computational Fluid Mechanics, 6*,
248–270. http://dx.doi.org/10.1080/19942060.2012.11
015419

Wu, Y., & Bakar, M. Z. A. (2000). Control of smoke
flow in tunnel fires using longitudinal ventilation
systems—A study of the critical velocity. *Fire Safety
Journal, 35*, 363–390. http://dx.doi.org/10.1016/
S0379-7112(00)00031-X

2

Designing robust feedback linearisation controllers using imperfect dynamic models and sensor feedback

Yongjing Wang[1]*, Lu Wang[2] and Mietek A. Brdys[3,4]

*Corresponding author: Yongjing Wang, Department of Mechanical Engineering, School of Engineering, University of Birmingham, Birmingham, UK
E-mail: yxw181@bham.ac.uk
Reviewing editor: Duc Pham, University of Birmingham, UK

Abstract: The paper considers key limitation of the feedback linearisation controller designed for nonlinear systems based on the imperfect nominal dynamics model and sensors. The model-reality differences cause signal leakages in the feedback linearised dynamics. As the leakages are the functions of the process variables, the resulting overall dynamics are again nonlinear with strong additive nonlinearities and the expected decoupling of the system dynamics is missing. In the paper, instead of using advanced control tools, we prove the robustness of the feedback linearisation method can also be significantly enhanced by employing several simple and classical methods cooperatively. For clear description and explanation, the methodology was illustrated based on a two-link manipulator case study, a classical multi-input multi-output coupled nonlinear system. The methods have genetic potential so that they can be applicable to a variety of case study systems and also further developed to become general methodologies.

Subjects: Engineering & Technology; Mechanical Engineering; Mechanical Engineering Design

Keywords: robotic manipulator; robust design; uncertainty; signal leakages; imperfect dynamic models; stiction friction

ABOUT THE AUTHORS

Yongjing Wang is a PhD researcher at the University of Birmingham. He has a wide interest in intelligent systems and structures.

Lu Wang obtained her BEng degree from the University of Birmingham and Huazhong University of Science and Technology in 2013. She is currently pursuing her PhD degree at the University of Southampton. Her research interest includes the optimisation and control of smart grid, vehicle-to-grid, energy storage and renewable energy integration.

Mietek A. Brdys, PhD, DSc, CEng, SMIEEE, FIMA (24/12/1946–25/07/2015) was Head of Control and Decision Support Research Laboratory, the University of Birmingham, and the founder of the Department of Control Systems Engineering, Gdansk University of Technology (Poland). He authored and co-authored over 220 refereed papers and 6 books. His research included intelligent decision support and control of large-scale complex systems, robust monitoring and control, and control and security of critical infrastructure systems.

PUBLIC INTEREST STATEMENT

In many applications, it is possible to use linear equations to describe the behaviour of the system accurately enough and linear controllers could work well enough if the system operates near the equilibrium. However, if it deviates too far from the equilibrium, the linear equations are no longer valid and linear controllers will fail. Feedback linearisation is a common approach used in controlling nonlinear systems. However, the control performance heavily depends on the accuracy of nominal models and sensors, which are possible to vary due to errors and wears in a long-term usage. Such uncertainties set a limit of the implementation of feedback linearisation design. Here, we present an approach to improve the robustness of a feedback linearisation controller by collaborating simple and classical tools to overcome a wide range of uncertainties in nominal models and sensors.

1. Introduction

Feedback linearisation control (Slotine & Li, 1991) has been widely used in controlling nonlinear and coupled multi-input multi-output (MIMO) systems (Focchi et al., 2010; Liu, Luo, & Rashid, 2000; Spong, Hutchinson, & Vidyasagar, 2005; Tahir, Iqbal, & Mustafa, 2009; Wen & Fang, 2012). Designing a feedback linearisation controller achieves the desired decoupling and linearisation only if the model is an accurate representation of reality. The model-reality differences cause signal leakages in the feedback linearised dynamics. As the leakages are the functions of process variables, the resulting overall dynamics are again nonlinear with strong additive nonlinearities. In addition, the expected decoupling of the system dynamics is missing and the resulting feedback linearisation controller is not robust enough to reach the desired performance. Such uncertainties are inevitable due to errors and wear in long-term usage.

Several tools have been developed to overcome uncertainties. Achieving robust decentralised architecture under strong link-reactions in robotic manipulators by applying the celebrated robust control technologies such as H-infinity is highly non-trivial. For the unlimited uncertainty, a number of adaptive control methods have been established for designing feedback linearisation controllers. A generic adaptive feedback linearisation controller with dynamic neural networks was developed for the control of uncertain nonlinear systems under no measurable states (Kulawski & Brdyś, 2000) and applied to high dynamic performance induction motor control (Focchi et al., 2010). The application of simple static neural networks to design adaptive controllers for manipulators under full measurement noise-free access to the needed variables was reported (Moradi & Malekizade, 2013). The generic adaptive feedback linearisation controller with Takagi-Sugeno fuzzy models adapted online was developed (Qi & Brdys, 2008, 2009). The adaptive linearisation feedback controller was developed for two-link robot arm under varying load torque and exact mathematical model (Yi & Chung, 1997). Global asymptotic stability of an adaptive output feedback tracking controller was developed for motion control of robot manipulator (Loría, Kelly, & Teel, 2005; Yarza, Santibanez, & Moreno-Valenzuela, 2013).

However, robust controllers are not necessarily built on any fancy tools and concepts. They can also be achieved by combining simple, classical and well-established tools. In this paper, instead of presenting new tools, we present a general approach, or a methodology, to develop robust feedback linearisation controllers based on accessible basic and popular control methods widely used. Depending on the radius of the uncertainty, two approaches are employed in the paper to enhance the control performance: robust approach and adaptive approach. For clear description and explanation, the methodology was illustrated based on a two-link manipulator case study, a typical MIMO coupled nonlinear system, which can be controlled by using various tools including model predictive control (Poignet & Gautier, 2000) and slide mode control (Ferrara & Magnani, 2007; He, Lin, & Wang, 2016; Mien, Kang, & Shin, 2014; Van, Kang, & Shin, 2013; Yu, Yu, Shirinzadeh, & Man, 2005). The approach described in this paper shows that robust control performance can also be achieved by using feedback linearisation even when big errors exist in nominal models and sensor feedback. We believe the strategy can be applied to a variety of systems and also further developed to become general methodologies.

The paper is organised as follows. Section 2 presents the mathematical model of a two-link robot arm and illustrates the feedback linearisation controller. Section 3 demonstrates a robust control approach for uncertain model parameters under a small uncertainty radius. When the uncertainty is significant due to large parameter errors and an existence of structural uncertainties in the nonlinear plant model dynamics, an adaptive approach is described in Section 4. The structural uncertainty means parts of the real model are missing in the nominal model.

2. Feedback linearisation control

2.1. Model of manipulator dynamic

The mechanical structure of a two-link robot arm, or two-link manipulator is shown in Figure 1. The mass m_1 and m_2 of each two links are uniformly distributed along their length L_1 and L_2. The torque is applied to the two joints to drive the movement of each link. The dynamics model of a two-link robot arm shown as Equations (1–4) is derived using the Lagrange method.

$$M(\theta)\ddot{\theta} + C(\theta, \dot{\theta})\dot{\theta} + G(\theta) + F(\dot{\theta}) + \tau_d = \tau \tag{1}$$

$$M(\theta) = \begin{bmatrix} m_1 L_1^2 + m_2 L_2^2 + m_2 L_1^2 + 2 m_2 L_1 L \cos\theta_2 & m_2 L_2^2 + m_2 L_1 L_2 \cos\theta_2 \\ m_2 L_2^2 + m_2 L_1 L_2 \cos\theta_2 & m_2 L_2^2 \end{bmatrix} \tag{2}$$

$$C(\theta, \dot{\theta}) = \begin{bmatrix} -2\dot{\theta}_2 m_2 L_1 L_2 \sin\theta_2 & -\dot{\theta}_2 m_2 L_1 L_2 \sin\theta_2 \\ \dot{\theta}_1 m_2 L_1 L \sin\theta_2 & 0 \end{bmatrix} \tag{3}$$

$$G(\theta) = \begin{bmatrix} m_1 L_1 g \cos\theta_1 + m_2 L_1 g \cos\theta_1 + m_2 L_2 g \cos(\theta_1 + \theta_2) \\ m_2 L_2 g \cos(\theta_1 + \theta_2) \end{bmatrix} \tag{4}$$

where L_1, L_2-lengths, m_1, m_2-masses, g-acceleration of gravity, $F(\dot{\theta})$-friction torques, τ_d-disturbance torques, τ_1, τ_2-control torques. The torques τ_1, τ_2 are the control inputs while the link positions θ_1, θ_2 are the outputs of the system. This is a classical MIMO coupled nonlinear system.

2.2. Effects of uncertainties on control performance of feedback linearisation

When coupling the feedback linearisation mapping (FLM) to the plant, the acceleration a becomes the control inputs, producing the feedback linearised plant of the second-order decoupled dynamics, as illustrated in Figure 2.

The design of the FLM requires exact knowledge of the plant dynamics model (mappings $M(\theta), C(\theta, \dot{\theta}), G(\theta), F(\dot{\theta})$), disturbance inputs τ_d and noise-free plant variable values $\theta(t), \dot{\theta}(t)$. However, in reality only the approximate models $\hat{M}(\theta), \hat{C}(\theta, \dot{\theta}), \hat{G}(\theta), \hat{F}(\dot{\theta})$ of $M(\theta), C(\theta, \dot{\theta}), G(\theta), F(\dot{\theta})$ are available. These discrepancies will produce additive leakage signals (LS) in the feedback linearised plant dynamics, which can be nonlinear functions of the plant variables. The state space model of the manipulator becomes:

$$\begin{aligned} \dot{x}_1 &= x_2 & x_1 &= \theta \\ & \quad \text{where} & & \\ \dot{x}_2 &= a - LS & x_2 &= \dot{\theta} \end{aligned} \tag{5}$$

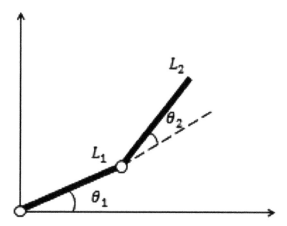

Figure 1. Two-link manipulator model.

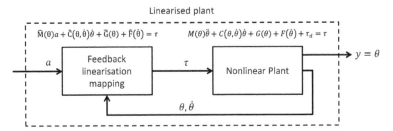

Figure 2. Linearised plant.

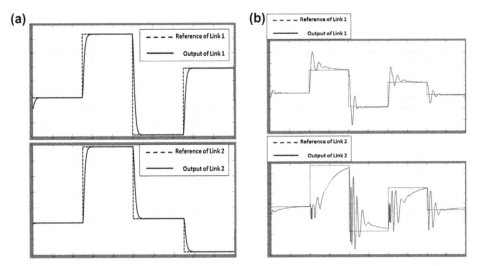

Figure 3. Effect of signal leakages in the feedback linearised dynamics. Control performance of (a) decentralised PID controller + Perfect FLM and (b) decentralised PID controller + imperfect FLM.

The leakage term can be derived accurately from the system equations, as shown in Equation (6), when the uncertainty is only in the disturbance torques.

$$LS(t) = M^{-1}(\tau_d(t) - \tau(t)) \tag{6}$$

The linearised plant can be controlled by using PID controllers in the outer position loops. The control outcome of a two-link manipulator with $L_1 = 20, L_2 = 15, m_1 = 0.7, m_2 = 0.5$ is shown in Figure 3(a), when the P, I, D terms equalled 10, 3, 0, respectively. However, good control performance is achievable only if the model is complete and parameters in the model are accurate. When the perfect model of the manipulator and noise-free measurement are not available, the error between the real parameters and the estimated parameters introduces nonlinear and coupled leakages, as shown in Figure 3(b) when parameter errors are 50% (Real: $L_1 = 20, L_2 = 15, m_1 = 0.7, m_2 = 0.5$; Estimated: $L_1 = 10, L_2 = 7.5, m_1 = 0.35, m_2 = 0.25$).

The result indicates the impact of nonlinear and coupled leakages caused by systematic uncertainty on feedback linearisation controllers. In this study, we present two approaches employing simple control tools in a combined manner to form new configurations in the controllers to enhance the robustness of controllers based on feedback linearisation.

3. Robust control approach

Cascaded-loop controller design is a classic design to overcome nonlinear and coupled leakages through high gains applied in the inner velocity loop. Regarding the two-link manipulator case study, this design is able to provide satisfactory control performance when errors are in the radius of 100%, described in Section 3.1. However, the high gains increase system's sensitivity to measurement noise, resulting in large position tracking error. To keep the functionality of the cascaded-loop controller, we propose to use a Kalman filter to eliminate the noise so that the whole control system is able to robustly achieve required tracking accuracy, described in Section 3.2.

3.1. Parameter uncertainties

Cascaded-loop design was used to overcome parameter uncertainties. Equation (6) shows that the LS caused by the parameter uncertainty appears entirely in the velocity loop. After implementing cascaded-loop configuration, the leakage introduced by the mechanical parameter uncertainty is compensated in the velocity loop or inner loop by using PI controller. While the outer loop provides a fast response and a high tracking accuracy assuming perfect performance of the inner velocity loop. In the inner loop, the P, I term in the two decentralised controllers are 30, 13 and 35, 20, respectively. The outer loop for each link is designed to employ a derivative feed-forward controller and a proportional controller with gain K_p.

The design of the outer loop controller starts from Equation (7):

$$\int (\dot{\theta}^{ref} + (\theta^{ref} - \theta)K_p) = \theta \tag{7}$$

Differentiating Equation (7) and rearranging the obtained terms yields the position error dynamics mode as follows:

$$\dot{\theta}^{ref} + \left(\theta^{ref} - \theta\right)K_p = \dot{\theta} \tag{8}$$

$$\dot{E} = \dot{\theta}^{ref} - \dot{\theta} = -K_p E \tag{9}$$

$$\therefore E = \exp(-K_p \cdot t) \tag{10}$$

From Equation (10), by using a proportional controller and derivative feed forward controller in the outer loop, the position reference tracking error converges to zero regardless of the initial error. The convergence rate can be arbitrarily fast by suitable choice of the gain K_p.

3.2. Noise

For FLM, the velocity and position measurement is used to compute desired torques. Noisy measurement introduces a large error to the computation results and control performance, especially when high gains are used in the inner loop.

A Kalman filter can be used to eliminate the noise. In our study, the acceleration of each link was measured by using acceleration sensors to predict the positions and velocities. The positions were also directly measured by using position sensors. According to the regular form of the continuous-time Kalman filter, the following equations are derived.

$$\dot{x} = Ax + Bu + m \tag{11}$$

$$y = Cx + n \tag{12}$$

$$p(m) \sim N(0, Q_c) \tag{13}$$

$$p(n) \sim N(0, R_c) \tag{14}$$

The state vector and corresponding matrices can be set in the following form:

$$x = \begin{bmatrix} x_1 \\ x_2 \end{bmatrix} = \begin{bmatrix} \text{position} \\ \text{velocity} \end{bmatrix} \tag{15}$$

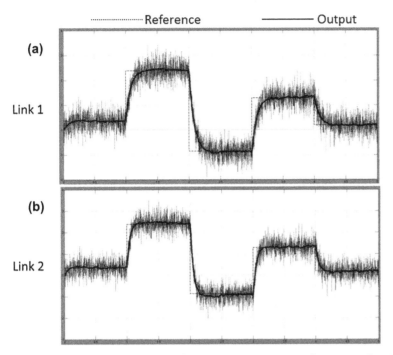

Figure 4. (a) Control performance of two separated PID controller + Imperfect FLM + Double cascaded loop controller for piecewise constant input signal and piecewise constant disturbance with noisy measurement; (b) Control performance of the designed robust controller.

$$A = \begin{bmatrix} 0 & 1 \\ 0 & 0 \end{bmatrix}, B = \begin{bmatrix} 0 \\ 1 \end{bmatrix}, C = [\ 1 \quad 0\ ] \tag{16}$$

Assume the noisy measurement of positions follows Gaussian distribution and the power of the reference signal is 1 and variance of noise m and n can be as large as 0.1.

$$Q_c = \begin{bmatrix} 0 \\ \sqrt{0.1} \end{bmatrix} \begin{bmatrix} 0 & \sqrt{0.1} \end{bmatrix} = \begin{bmatrix} 0 & 0 \\ 0 & 0.1 \end{bmatrix} \tag{17}$$

$$R = 0.1 \tag{18}$$

The estimated result is

$$\hat{\dot{x}} = A\hat{x} + Bu + K(y - C\hat{x}) \tag{19}$$

The Kalman filter gains are updated as Equations (21) and (22):

$$\dot{P} = -PC^T R_c^{-1} CP + AP + PA^T + Q_c \tag{20}$$

$$K = PC^T R_c^{-1} \tag{21}$$

When the system operates under a piecewise constant disturbance of high frequency and a noise with SNE = 1 dB and the parameter errors are 50%, the tracking results are shown in Figure 4. The newly designed controller achieves an average tracking error = 0.5%, overshoot = 0% and settling time = 0.01 s. This design minimised the effects of noise due to suitable high gains in cascaded-loop configurations, and increase the robustness of the system.

Overall, the robustness of feedback linearisation controller can be improved by using classical cascaded-loop configuration without amplifying the effects of sensor noise due to high gains. The design could provide satisfactory control performance when parameter errors are in the radius of 100%.

4. Adaptive control approach

The robust design can ensure a good control performance when the parameter errors stay in the radius of 100%. However, when uncertainty exists in a larger radius, such as large parameter errors over 100% and incomplete dynamic models (or structural uncertainty), the previous robust control approach is not strong enough. In terms of two-link manipulator dynamics, the stiction friction is missing in the original dynamic models (1–4) and introduces structural uncertainty. In this section, we present a simple adaptive controller adopting a number of simple and basic control tools and algorithms to correct parameters online (Section 4.1) and overcome unknown structural uncertainty (Section 4.2) simultaneously.

4.1. Large parameter errors

A sub-system called a parameter estimator is added to the system in order to update the current parameter values based on the comparison of the real and predicted/estimated outputs. The parameter update algorithm aims at forcing the output mismatch to zero. Figure 5 describes the new configuration of the system.

Considering the relationship between the plant and the FLM, the following equations are derived from Equations (1–5).

$$M^{-1}(\theta)\left[\widehat{M}(\theta)a + \widehat{C}(\theta,\dot{\theta})\dot{\theta} + \widehat{G}(\theta) + \widehat{F}(\dot{\theta})\right] - F(\dot{\theta}) - G(\theta) - C(\theta,\dot{\theta})\dot{\theta} = \ddot{\theta} \tag{22}$$

$$\left(\widehat{M} - M\right)\ddot{\theta} + \widehat{M}(a - \ddot{\theta}) + \left(\widehat{C} - C\right)\dot{\theta} + \left(\widehat{G} - G\right) = 0 \tag{23}$$

$$\Delta M\ddot{\theta} + \widehat{M}e + \Delta C\dot{\theta} + \Delta G = 0 \tag{24}$$

In Equation (24), e is the error between the measured acceleration and the expected acceleration into the FLM while Δ denotes the parameter errors on the values of the real and nominal model mappings. The aim of parameter estimator is to solve the equation $\Delta M\ddot{\theta} + \widehat{M}e + \Delta C\dot{\theta} + \Delta G = 0$ and generate new parameter values to update the old ones.

In Equation (24)

$$\Delta M = \widehat{M} - M = \begin{bmatrix} \Delta(m_1L_1^2) + \Delta(m_2L_2^2) + \Delta(m_2L_1^2) + 2\Delta(m_2L_1L_2)\cos\theta_2 & \Delta(m_2L_2^2) + \Delta(m_2L_1L_2)\cos\theta_2 \\ \Delta(m_2L_2^2) + \Delta(m_2L_1L_2)\cos\theta_2 & \Delta(m_2L_2^2) \end{bmatrix} \tag{25}$$

$$\Delta C = \widehat{C} - C = \begin{bmatrix} -2\Delta(m_2L_1L_2)\dot{\theta}_2\sin\theta_2 & -\Delta(m_2L_1L_2)\dot{\theta}_2\sin\theta_2 \\ \Delta(m_2L_1L_2)\dot{\theta}_1\sin\theta_2 & 0 \end{bmatrix} \tag{26}$$

$$\Delta G = \widehat{G} - G = \begin{bmatrix} \Delta(m_1L_1)g\cos\theta_1 + \Delta(m_2L_1)g\cos\theta_1 + \Delta(m_2L_2)g\cos(\theta_1 + \theta_2) \\ \Delta(m_2L_2)g\cos(\theta_1 + \theta_2) \end{bmatrix} \tag{27}$$

Assuming that

$$X = \begin{bmatrix} X_1 \\ X_2 \\ X_3 \\ X_4 \\ X_5 \\ X_6 \\ X_7 \end{bmatrix} = \begin{bmatrix} \Delta(m_1L_1^2) \\ \Delta(m_2L_1^2) \\ \Delta(m_2L_2^2) \\ \Delta(m_2L_1L_2) \\ \Delta(m_1L_1) \\ \Delta(m_2L_1) \\ \Delta(m_2L_2) \end{bmatrix} \tag{28}$$

The Equations (25–27) can be transformed to (29–31)

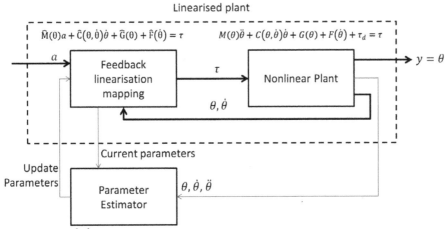

Figure 5. Block diagram of the control system containing a parameter estimator.

Table 1. Ultra-large error								
Parameter error: 1,000%	Feedback linearisation mapping				Real dynamic model			
	m_1	m_2	L_1	L_2	m_1	m_2	L_1	L_2
Large parameters	20	15	0.7	0.5	200	150	7	5
Small parameters	20	15	0.7	0.5	2	1.5	0.07	0.05

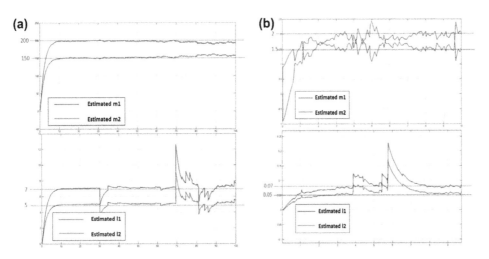

Figure 6. Estimated values of $m1$, $m2$, $L1$ and $L2$: (a) ultra-large parameters; (b) ultra-small parameters.

$$\Delta M = \begin{bmatrix} X_1 + X_2 + X_3 + 2X_4\cos\theta_2 & X_3 + X_4\cos\theta_2 \\ X_3 + X_4\cos\theta_2 & X_3 \end{bmatrix} \qquad (29)$$

$$\Delta C = \begin{bmatrix} -2X_4\dot{\theta}_2\sin\theta_2 & -X_4\dot{\theta}_2\sin\theta_2 \\ X_4\dot{\theta}_1\sin\theta_2 & 0 \end{bmatrix} \qquad (30)$$

$$\Delta G = \begin{bmatrix} X_5 g\cos\theta_1 + X_6 g\cos\theta_1 + X_7 g\cos(\theta_1 + \theta_2) \\ X_7 g\cos(\theta_1 + \theta_2) \end{bmatrix} \qquad (31)$$

Thus, the whole function (24) can now be transformed to (32)

$$K(\ddot{\theta},\dot{\theta},\theta)X + \hat{M}\left(\hat{m}_1,\hat{m}_2,\hat{L}_1,\hat{L}_2\right)e = 0 \tag{32}$$

where

$$K = \begin{bmatrix} \ddot{\theta}_1 & \ddot{\theta}_1 & \ddot{\theta}_1+\ddot{\theta}_2 & (2\ddot{\theta}_1+\ddot{\theta}_2)\cos\theta_2 - 2\dot{\theta}_1\dot{\theta}_2\sin\theta_2 - \dot{\theta}_2^2\sin\theta_2 & g\cos\theta_1 & g\cos\theta_1 & g\cos(\theta_1+\theta_2) \\ 0 & 0 & \ddot{\theta}_1+\ddot{\theta}_2 & \ddot{\theta}_1\cos\theta_2 + \dot{\theta}_1^2\sin\theta_2 & 0 & 0 & g\cos(\theta_1+\theta_2) \end{bmatrix} \tag{33}$$

$$\hat{M}(\theta) = \begin{bmatrix} \hat{m}_1\hat{L}_1^2 + \hat{m}_2\hat{L}_2^2 + \hat{m}_2\hat{L}_1^2 + 2\hat{m}_2\hat{L}_1\hat{L}_2\cos\theta_2 & \hat{m}_2\hat{L}_2^2 + \hat{m}_2\hat{L}_1\hat{L}_2\cos\theta_2 \\ \hat{m}_2\hat{L}_2^2 + \hat{m}_2\hat{L}_1\hat{L}_2\cos\theta_2 & \hat{m}_2\hat{L}_2^2 \end{bmatrix} \tag{34}$$

$$e = \begin{bmatrix} e_1 \\ e_2 \end{bmatrix} = \begin{bmatrix} a_1 - \ddot{\theta}_1 \\ a_2 - \ddot{\theta}_2 \end{bmatrix} \tag{35}$$

Equation (33) is valid at any time instant. In order to calculate parameters L_1, L_2, m_1 and m_2, the data is sampled in a time stream at a different time instant so that K is expanded to K_t and thus Equation (32) is transformed into Equation (36).

$$X_t = -K_t^{-1}(\ddot{\theta},\dot{\theta},\theta)\hat{M}_t(\hat{m}_1,\hat{m}_2,\hat{L}_1,\hat{L}_2)e_t \tag{36}$$

when sampling three times,

$$X_t = \begin{bmatrix} X_{t1} \\ X_{t2} \\ X_{t3} \end{bmatrix} = -\begin{bmatrix} K_{t1} \\ K_{t2} \\ K_{t3} \end{bmatrix}^{-1} \begin{bmatrix} \hat{M}_{t1} \\ \hat{M}_{t2} \\ \hat{M}_{t3} \end{bmatrix} \tag{37}$$

There are three different time instants, $t1, t2, t3$, that K, X and M sampled at. As seen in Equation (33), K at different time t is always following a certain proportional law. Thus the maximum rank of K_t is 5 and there is infinite number of roots for X. K_t can be simplified and transformed into Equation (38):

$$K_t = \begin{bmatrix} 1 & 1 & 0 & 0 & 0 & 0 & 0 \\ 0 & 0 & 1 & 0 & 0 & 0 & 0 \\ 0 & 0 & 0 & 1 & 0 & 0 & 0 \\ 0 & 0 & 0 & 0 & 1 & 1 & 0 \\ 0 & 0 & 0 & 0 & 0 & 0 & 1 \\ 0 & 0 & 0 & 0 & 0 & 0 & 0 \end{bmatrix} \tag{38}$$

The pseudo-inverse operator is a tool to solve this problem. The estimated results of X_3, X_4 and X_7 are accurate. Therefore, assuming that the mass of the whole manipulator is available, the value of parameters can be calculated online by using the equations below:

$$\left.\begin{array}{r} \hat{m}_2\hat{L}_2^2 - X_3 = m_2L_2^2 = Y_1 \\ \hat{m}_2\hat{L}_1\hat{L}_2 - X_4 = m_2L_1L_2 = Y_2 \\ \hat{m}_2\hat{L}_2 - X_7 = m_2L_2 = Y_3 \\ \dfrac{Y_1}{Y_3} = L_2 \\ \dfrac{Y_2}{Y_3} = L_1 \\ \dfrac{Y_3^2}{Y_1} = m_2 \\ mass - m_2 = m_1 \end{array}\right\} \tag{39}$$

To test the algorithm, it is assumed that the initial parameters in the FLM are fixed and two parameter data-sets are used in the real dynamic models: extremely large values and extremely small values, as shown in Table 1. The performance of the parameter estimator using the algorithm described above is given in Figure 6.

When the real parameters are 10 times larger or smaller than the parameters in the FLM, the estimated parameter fluctuates approximately within a 10% range around the target value. The parameter-learning algorithm is effective since it reduces the parameter error from 1,000% to around 10%, which are in the working radius of the robust controller.

4.2. Structural uncertainties

According to the stiction phenomena, when the robot arm starts to move and the velocity is small, it experiences high torque at its joints. Thus, stiction friction depends on the velocity in a nonlinear way. However, this part is missing in the nominal model, causing structural uncertainties. The relationship between the stiction friction and velocity in the steady state is shown in Figure 7.

The stiction friction torque can be expressed by an exponential function shown as follows:

$$F_s(v) = f_s \cdot \exp(-\frac{v^2}{v_s}) \tag{40}$$

In Equation (40), f_s is the maximum value of the stiction friction and v_s is a parameter set to be a very small value. Stiction frictions F_{s1} of link 1 and F_{s2} of link 2 are indicated by Equations (41) and (42), respectively.

$$F_{s1}(\dot{\theta}_1) = f_s \cdot \exp(-\frac{\dot{\theta}_1^2}{v_s}) \cdot sign(\dot{\theta}_1) \tag{41}$$

$$F_{s2}(\dot{\theta}_2) = f_s \cdot \exp(-\frac{\dot{\theta}_2^2}{v_s}) \cdot sign(\dot{\theta}_2) \tag{42}$$

The overall friction matrix F in (1) is expanded to Equation (43):

$$F(\dot{\theta}) = \begin{pmatrix} f_{11}\dot{\theta}_1 + f_{12}sign(\dot{\theta}_1) + f_s \cdot \exp(-\frac{\dot{\theta}_1^2}{v_s}) \cdot sign(\dot{\theta}_1) \\ f_{21}\dot{\theta}_2 + f_{22}sign(\dot{\theta}_2) + f_s \cdot \exp(-\frac{\dot{\theta}_2^2}{v_s}) \cdot sign(\dot{\theta}_2) \end{pmatrix} \tag{43}$$

The stiction torque expressed by Equation (43) constitutes a real model of the friction. The leakage signal in the dynamical model (6) of the feedback linearised system equals to:

$$LS = M^{-1}F_s(\dot{\theta}) \tag{44}$$

where inertia matrix $M = \begin{bmatrix} m_{11} & m_{12} \\ m_{21} & m_{22} \end{bmatrix}$ with none of its components zero, and $F_s = \begin{bmatrix} F_{s1} \\ F_{s2} \end{bmatrix}$,

where F_{s1} and F_{s2} are defined in Equations (41) and (42), respectively. Hence,

$$M^{-1}F_s(\dot{\theta}) = \frac{1}{|M|} \begin{pmatrix} m_{22}F_{s1}(\dot{\theta}_1) - m_{12}F_{s2}(\dot{\theta}_2) \\ -m_{21}F_{s1}(\dot{\theta}_1) + m_{11}F_{s2}(\dot{\theta}_2) \end{pmatrix} \tag{45}$$

Therefore, for each link, the leakage into the link due to the uncertainty in the stiction friction varies with velocities of both links in a nonlinear manner. Simulation results (Figure 8) of the system with integral controllers under the stiction friction illustrate the impact of static friction on the tracking performance.

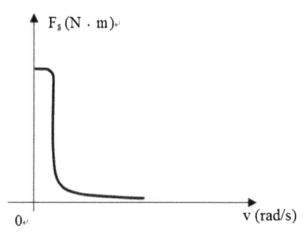

Figure 7. Continuously changing approximation of relationship between static torque and velocity

This structural uncertainty can be compensated simply by using a classical gain-scheduling mechanism. The linguistic setting details of the gain scheduling are illustrated in Figure 9.

As the stiction leakages directly appear in the inner loops, the gain scheduling is only applied to the PI controllers in the inner loops. This helps to stop/reduce the impacts transferred to the outer position loops. Taking into account that the actuator is unable to generate rapidly changing torques, soft gain scheduling is used, where changes of the gain values are distributed over time as opposed to the hard switching of the gains.

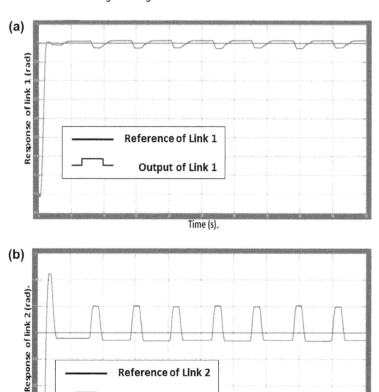

Figure 8. Impact of stiction friction that missed in the nominal model.

The soft gain scheduling is shown in Equation (46).

$$GS_i(\dot{\theta}) = \alpha_{i,1}\exp(-\dot{\theta}_1^2/v_s) + \alpha_{i,2}\exp(-\dot{\theta}_2^2/v_s) + 1 \qquad (46)$$

where $i = 1, 2$ denotes the link number.

The control performance is given in Figure 10, by using the parameter estimator to correct parameters and the gain scheduling to compensate structural uncertainties (stiction friction). Due to the inner loop including a gain-scheduled PI controller, the control system is able to overcome not only static torque, but also piecewise-constant disturbance due to the integral fast controller. When the parameter error is as large as 1,000% and structural uncertainty exists in the form of stiction friction, it is demonstrated that the steady-state error of link 1 lower than 0.01% and the steady-state error of link 2 lower than 0.03%.

Overall, the robustness of feedback linearisation controller can be improved by using a simple parameter learning algorithm and a classical gain-scheduling mechanism. When parameter errors are 1,000%, this new and simple approach is capable of achieving a good tracking performance.

5. Conclusion
This paper proves that instead of using a unique advanced tool, the robustness of feedback linearisation controller can be significantly improved by a combination of simple and basic tools. This general

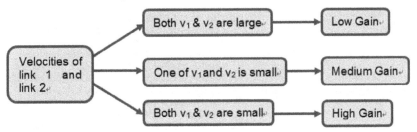

Figure 9. Linguistic setting details of gain scheduling.

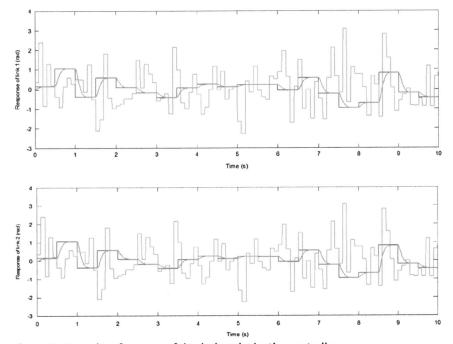

Figure 10. Control performance of the designed adaptive controller.
Notes: Actual outputs (red); reference outputs (blue); disturbance (green).

approach, or methodology, is able to overcome significant uncertainties in nominal models and sensors. The methodology was presented based on a two-link manipulator case study. When uncertainty stays in a small radius (parameter error < 100%), it is proposed to employ a robust control approach, in which cascaded-loop design and Kalman filter are used cooperatively, as the former is able to overcome parameter errors through high gains in the inner loop, while the latter prevents amplifying the noise owing to such high gains. When uncertainty is in a large radius, it is proposed to use an adaptive control approach consisting of a simple parameter learning algorithm and gain-scheduling mechanism to correct parameters and eliminate structural uncertainties. As different tools dealing with different aspects of uncertainties, the technique can be further supplemented and tailored to fit requirements, and become a general methodology.

Funding
The authors received no direct funding for this research.

Author details
Yongjing Wang[1]
E-mail: yxw181@bham.ac.uk
ORCID ID: http://orcid.org/0000-0002-9640-0871
Lu Wang[2]
E-mail: lw5g13@soton.ac.uk
Mietek A. Brdys[3,4]
E-mail: m.brdys@bham.ac.uk
[1] Department of Mechanical Engineering, School of Engineering, University of Birmingham, Birmingham, UK.
[2] Faculty of Engineering and Environment, University of Southampton, Southampton, UK.
[3] Department of Electronic, Electrical and Systems Engineering, School of Engineering, University of Birmingham, Birmingham, UK.
[4] Faculty of Electrical and Control Engineering, Gdansk University of Technology, Gdansk 80-952, Poland.

References
Ferrara, A., & Magnani, L. (2007). Motion control of rigid robot manipulators via first and second order sliding modes. *Journal of Intelligent and Robotic Systems, 48*, 23–36.

Focchi, M., Guglielmino, E., Semini, C., Boaventura, T., Yang, Y. S., & Caldwell, D. G. (2010). Control of a hydraulically-actuated quadruped robot leg. *2010 IEEE International Conference on Robotics and Automation* (pp. 4182–4188). Anchorage, AK.

He, S., Lin, D., & Wang, J. (2016, March). Chattering-free adaptive fast convergent terminal sliding mode controllers for position tracking of robotic manipulators. *Proceedings of the Institution of Mechanical Engineers, Part C: Journal of Mechanical Engineering Science, 230*, 514–526.

Kulawski, G. J., & Brdyś, M. A. (2000). Stable adaptive control with recurrent networks. *Automatica, 36*, 5–22. http://dx.doi.org/10.1016/S0005-1098(99)00092-8

Liu, Z. Z., Luo, F. L., & Rashid, M. H. (2000). Nonlinear load-adaptive MIMO controller for high performance DC motor field weakening. In *Conference and Proceedings on Power Engineering Society Winter Meeting, 2000 IEEE* (Vols. 1–4, pp. 332–337). Singapore.

Loría, A., Kelly, R., & Teel, A. R. (2005). Uniform parametric convergence in the adaptive control of mechanical systems. *European Journal of Control, 11*, 87–100. http://dx.doi.org/10.3166/ejc.11.87-100

Mien, V., Kang, H., & Shin, K. (2014). Adaptive fuzzy quasi-continuous high-order sliding mode controller for output feedback tracking control of robot manipulators. *Proceedings of the Institution of Mechanical Engineers, Part C: Journal of Mechanical Engineering Science, 228*, 90–107.

Moradi, M., & Malekizade, H. (2013). Neural network identification based multivariable feedback linearization robust control for a two-link manipulator. *Journal of Intelligent & Robotic Systems, 72*, 167–178.

Poignet, P., & Gautier, M. (2000). Nonlinear model predictive control of a robot manipulator. In *Proceedings on Advanced Motion Control, 2000* (pp. 401–406). 6th International Workshop on, Nagoya, Japan. doi:10.1109/AMC.2000.862901

Qi, R. Y., & Brdys, M. A. (2008). Stable indirect adaptive control based on discrete-time T-S fuzzy model. *Fuzzy Sets and Systems, 159*, 900–925. http://dx.doi.org/10.1016/j.fss.2007.08.009

Qi, R. Y., & Brdys, M. A. (2009). Indirect adaptive controller based on a self-structuring fuzzy system for nonlinear modeling and control. *International Journal of Applied Mathematics and Computer Science, 19*, 619–630.

Slotine, J.-J. E., & Li, W. (1991). *Applied nonlinear control* (Vol. 199). Englewood Cliffs, NJ: Prentice-Hall.

Spong, M. W., Hutchinson, S., & Vidyasagar, S. (2005). *Robot modeling and control*. New York, NY: Wiley.

Tahir, F., Iqbal, N., & Mustafa, G., (2009). Control of a nonlinear coupled three tank system using feedback linearization. *2009 Third International Conference on Electrical Engineering* (pp. 1–6). Lahore.

Van, M., Kang, H., & Shin, K. (2013). Backstepping quasi-continuous high-order sliding mode control for a Takagi–Sugeno fuzzy system with an application for a two-link robot control. *Proceedings of the Institution of Mechanical Engineers, Part C: Journal of Mechanical Engineering Science.* Retrieved from http://pic.sagepub.com/content/early/2013/10/23/0954406213508936

Wen, T., & Fang, J. C. (2012). A feedback linearization control for the nonlinear 5-DOF flywheel suspended by the permanent magnet biased hybrid magnetic bearings. *Acta Astronautica, 79*, 131–139.

Yarza, A., Santibanez, V., & Moreno-Valenzuela, J. (2013). An adaptive output feedback motion tracking controller for robot manipulators: Uniform global asymptotic stability and experimentation. *International Journal of Applied Mathematics and Computer Science, 23*, 599–611.

Yi, S. Y., & Chung, M. J. (1997). A robust fuzzy logic controller for robot manipulators with uncertainties. *IEEE Transactions on Systems, Man, and Cybernetics, Part B: Cybernetics, 27*, 706–713.

Yu, S., Yu, X., Shirinzadeh, B., & Man, Z. (2005). Continuous finite-time control for robotic manipulators with terminal sliding mode. *Automatica, 41*, 1957–1964.

3

Augmentation of distillate yield in "V"-type inclined wick solar still with cotton gauze cooling under regenerative effect

P.U. Suneesh[1,2], John Paul[2], R. Jayaprakash[3]*, Sanjay Kumar[4] and David Denkenberger[5]

*Corresponding author: R. Jayaprakash, Department of Physics, Sri Ramakrishna Mission Vidhyalaya College of Arts and Science, Coimbatore 641020, India
E-mail: jprakash_jpr@rediffmail.com
Reviewing editor: Raya Al-Dadah, University of Birmingham, UK

Abstract: Water flowing over bare glass is not evenly distributed over the width of the glass cover. Effective temperature reduction in the glass cover is not possible in this case. Thin cotton gauze is used over the glass cover to solve this problem. The amount of water required to cool the glass cover is low and the flowing water is collected and fed into a hot water reservoir. Water from the hot water reservoir is directed to the wick absorber by a drip valve. The system was tested in two ways (i) regenerative effect with cotton gauze (ii) regenerative effect without cotton gauze. The inclined wick absorber ensures that the surface is always wet due to capillary action and there were no dry spots. Excess flow in the wick absorber is pumped back to the hot water reservoir. Due to low thermal inertia of the wick and regenerative effect, the system has quick start-up times, as well as higher operating temperatures. This resulted in higher distillate yield than simple solar stills. Since the hot water reservoir remained warm enough during the night hours, a reasonable amount of nocturnal distillate output was also obtained. Optimum inclination of the wick absorber was found to be 20° and mass flow rate through the wick absorber was 200 ml/minute. The system with cotton gauze produced distillate yield of 6,300 ml/m^2 whereas the system without cotton gauze produced distillate yield of 5,600 ml/m^2.

ABOUT THE AUTHOR

P.U. Suneesh is an assistant professor in the Department of Applied Science, MES College of Engineering, Kuttippuram, Kerala affiliated to Kerala Technological University. He received his MSc degree in Physics from Calicut University and MPhil degree in Solid State Physics from Bharathidasan University, Trichy, Tamilnadu. He has more than 10 years experience of teaching Physics to undergraduate engineering students. He has published eight research papers in reputed journals. His research interests include Renewable energy, Nano gas sensors, and Desalination. His research work is associated with Research and Development Centre, Solar Energy Laboratory, Department of Physics and Civil and Architectural Engineering, Tennessee State University, Tennessee, USA. The major goal of this work was to increase the distillate yield using locally available cheap/waste materials. South India is famous for textile industry and plenty of cotton waste materials are available.

PUBLIC INTEREST STATEMENT

A solar still converts salt/brackish water into potable water using solar radiation. The basin heats up due to radiation and eventually water in the basin/absorber evaporates and condenses on the glass cover. The temperature gradient between the evaporating surface and condensing surface plays an important role in the distillate yield. In this work, the temperature difference between the glass and the basin was increased by cooling the glass cover using water flow over the glass cover. For low flow rates (75 ml/min, 150 ml/min etc.), water flowing over the glass cover is not evenly distributed. This can be solved using very thin cotton gauze over the glass cover. For high flow rates, the flow will be turbulent and more amount of water is required to cool. The flowing water sorted in the hot water reservoir for regenerative effect.

Subjects: Environment & Agriculture; Mechanical Engineering; Physical Sciences

Keywords: solar still; desalination; V-type; wick; drip system; gauze

1. Introduction

A chronic drinking water shortage is one of the most important issues in the developing countries, and drinking water from dirty water sources causes serious damage to health, especially in infants, children, and elderly people. The lack of potable water is a big problem in many areas. Underground water, where exists, is sometimes salty and cannot be used for drinking. Solar stills can be used for low-capacity and self-reliant water supply systems, since they can produce drinking water by solar energy only, and do not need other energy sources such as fuel or electricity. Renewable energy sources have been used to supply thermal and/or electrical energies for desalination in remote areas. Among these sources is solar energy, which has gained increasing attention because of its abundance in locations suffering from lack of fresh water. Wick-type solar stills are widely accepted and are known for their high distillate yield. Wick-type stills show improved distillate yield compared to the conventional water-filled stills (Velmurugan, Gopalakrishnan, Raghu, & Srithar, 2000). A study reported on improved design of a single slope coupled to a wick in order to enhance still output. The energy absorbed by the absorber basin is mostly transferred to the water. In water filled systems, much of the morning solar energy is used for providing sensible heat so (Minasian & Al-Karaghouli, 1995) most of the water filled systems show high distillate yield after noon.

Performance of a tilted-wick-type solar still with the effect of water flowing over the glass has been studied (Janarthanan, Chandrasekaran, & Kumar, 2006) and the conclusion was that glass cover temperature decreases significantly due to the water flowing over the glass cover which causes fast internal evaporation during peak sunny hours. The authors recorded an optimum rate of water flow 1.5 m/s over the glass cover. Several approaches have been used to cool the glass cover and to utilize the latent heat of condensation released into the glass cover (Tiwari & Bapeshwara Rao, 1984; Tiwari, Madhuri, & Garg, 1985), cooling by flowing water film on top of the glass cover. The main focus was to maintain a temperature difference between basin water and condensing cover in order to get high distillate output. Different types of wick materials have been used (Kalidasa Murugavel & Srithar, 2011) on a basin-type solar still and it was concluded that using a thin cotton wick optimized output. Performance of a "V"-type solar still with a charcoal absorber has been studied by Selvakumar, Kumar, and Jayaprakash (2008) and its advantages included the center collection of all the condensation. An inclined solar water distillation system was designed and tested under actual environmental conditions of northern Cyprus, and the system produces fresh water and hot water simultaneously (Aybar, Egelioğlu, & Atikol, 2005). Experimental performance evaluation of a stand-alone point focus parabolic solar still was conducted by Gorjian, Ghobadian, Tavakkoli Hashjin, and Banakar (2014) with two plate heat exchangers to pre-heat the salt water. The maximum productivity of 5.12 kg within 7 h in a day was measured with the maximum average solar insolation of 626.8 W/m^2. Somwanshi and Tiwari (2014) reported that solar still output enhances by flowing water from an air cooler over the glass cover. It was seen that for different climatic zones in the Indian plain, annual yield can be increase between 41.3% and 56.5%. Moreover, the distillate output increases slightly with an increase in the mass flow rate and tends to saturate around 0.075 kg/s. A solar water heater coupled with a still was studied by Sampathkumar and Senthilkumar (2012). The work reported that the storage water temperature reached 60°C and increased the yield by 77% compared to the passive still. Sakthivel, Shanmugasundaram, and Alwarsamy (2010) conducted an experimental study on a regenerative solar still with an energy storage medium—jute cloth. The jute cloth was kept vertical in the middle of basin saline water and also attached to the rear wall of the still. It was found that cumulative yield in the regenerative still with jute cloth increases approximately by 20% and efficiency increases by 8% with low cost as the jute cloth is very cheap and easily available. Malaiyappan and Elumalai (2016) have reported improved nocturnal output using a helical copper coil, aluminum fins, a long hollow stainless steel tube, and an iron plate coupled with the basin. The solar still with various basin materials arranged in a lengthwise direction was more effective, compared with the one arranged breadthwise. Refalo, Ghirlando, and Abela

(2015) conducted experiments with a solar chimney and condenser coupled with solar still and found that the modified still generated 5.1 L/m² day with the majority of the yield (59%) condensing in the condensers of the still.

In the current work, an inclined wick-type solar still with water flowing over the top cover integrated with a hot water reservoir has been constructed. Unlike other solar stills with water flowing over the glass plate, the water flow over the glass here is made uniform by placing a thin layer of cotton gauze (Suneesh, Jayaprakash, Arunkumar, & Denkenberger, 2014). The cooling water is conveyed to a water reservoir which in turn is connected to the inclined wick absorber by a valve. The excess flow produced inside the still is pumped back to the water reservoir. This setup requires less amount of water to cool glass surface compared to other water flowing systems of its kind. So, a new attempt has been made using a regenerative effect for increasing yield rate due to supply of pre-heated water. Performance due to variation in mass flow rate inside the still and angle of inclination of wick absorber are also studied.

2. Materials and methods

A photographic view of the still is shown in Figure 1. Figure 2 shows the cotton gauze arrangement over the glass cover. Figure 3 shows a schematic diagram of solar still with a hot water reservoir. The length and breadth of wick absorber are 2 m and 0.75 m, respectively, and the area of wick absorber is 1.5 m². The wick absorber is inclined inward to the center of the still and the angle of inclination can be varied between 0° and 25°. Water fed into the wick absorber is spread under capillary action and no dry spots are produced. The excess water is pumped back to the water reservoir. An optimum inward slope is maintained for top cover, reported by Singh and Tiwari (2004) for India's climatic conditions. An arrangement of water flowing over the glass cover of the solar still was made with the help of PVC pipe of ½ inch diameter. The length of the pipe was taken exactly equal to the width of the still so that the water would not flow in the lengthwise direction. An external water tank is used and is connected to distribution pipes on either side of the still as shown in the Figure 1. A number of holes were made in the pipe spaced equally to produce uniform flow over the glass cover. After the water flows over the glass, it is collected at the center of the still by a water collection channel and fed into the reservoir. The photographic view of the still with collection channels for distillate and water flowing over the glass is shown in Figure 4. A rectangular water reservoir of 50 l was designed; the side and bottom area of the water reservoir is insulated to minimize the heat loss. A valve is used to control the water flow from the hot water reservoir to the still. The water then trickles down through the wick, creating a layer of water all over the wick absorber. Some of the water evaporates and condenses as it touches the cool glass cover. The condensate flows into a central condensate channel and is taken out of the still. The rest of the fed water (excess flow), which is hot water, flows into another collection channel and is pumped back to the reservoir. A 20 W suction pump was used to pump the excess water to the hot water reservoir. The saline water is replaced periodically so that

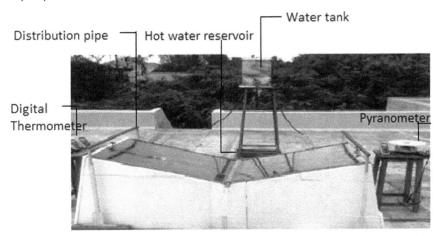

Figure 1. Photographic view of "V"-type solar still without cotton gauze under regenerative effect.

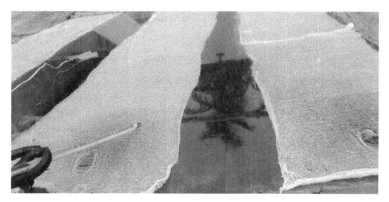

Figure 2. Cotton gauze over the glass cover.

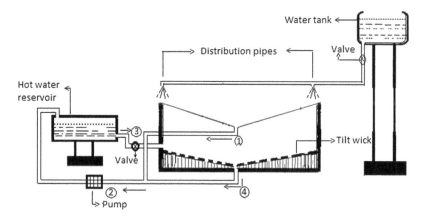

①Water flowing over north and south facing glass is collected ② Water is pumped to the hot water reservoir ③ Water flows from the reservoir to the tilt basin(flow under gravity) ④ Excess flow in the basin is pumped back to the reservoir

Figure 3. Schematic diagram of the solar still.

Figure 4. Photographic view of still showing collection channels for distillate and water flowing over glass.

no solids precipitate out. In the first mode of experiment (regenerative effect with cotton gauze), since cotton gauze spreads water uniformly over the glass cover, minimal water was used just to spread over the glass cover and the flow rate was 75 ml/min. The discharge amount per hour of water from the glass cover to the hot water reservoir was 4,500 ml (approximately). A 50 l hot water reservoir was sufficient to store the water for the 9 h (9.00 am–6.00 pm) of the experiment under regenerative effect. In the second mode of experiment, the still was tested without cotton gauze with same water flow rate over the glass cover.

   The experimental study conducted from 9.00 am to 6.00 pm. Nocturnal output was also recorded. The performance of the still due to parameters like effect of water mass flow rate through wick absorber and angle of inclination of wick absorber were experimentally tested in typical sunny days. Solar radiation, temperature of hot water, and water collection at the outlet position were recorded at regular intervals.

3. Results and discussion

Variation of solar radiation and ambient temperature during the experiment is shown in Figure 5. Average radiation received was 665 W/m² and the ambient air temperature during the sunny days was measured between 24 and 36°C. The still technical and operational parameters are shown in Table 1. Accuracies and error for various measuring instruments are given in Table 2. The comparison of glass temperature of Figure 6 indicates a range of 34–53°C without cotton gauze and 34–48°C with cotton gauze. The temperature is relatively high for the still with water flow without cotton gauze. The test for the 15° inclined wick absorber with cotton gauze shows the effects of mass flow rate on the fresh water generation. As seen in Figure 7, the freshwater generation increases when the mass flow increases and then decreases for further increase in the mass flow rate. The test for the 20° inclined wick absorber shows optimum results. The results were of course also affected by environmental conditions, such as changing solar intensity, ambient temperature, and wind. In this study, the tests were performed and tabulated for 8 h. Average temperature of the hot water for the day ranges from 44 to 51°C for the 15° wick absorber, 43 to 48°C for the 20° wick absorber, and 43 to 46°C for the 25° wick absorber (Figure 8). Results of distillate output show that the best output is 6,300 ml/m² for 200 ml/min mass flow rate at an inclined angle 20° of wick absorber. The still shows significant distillate yield rate even after sunset irrespective of tilts and flow parameters. This is mainly due to the hot water in the reservoir. The average hot water reservoir temperature during the

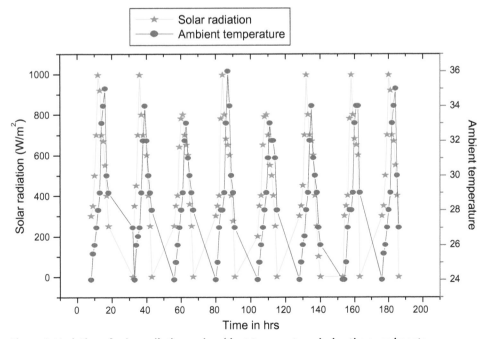

Figure 5. Variation of solar radiation and ambient temperature during the experiments.

Table 1. Still technical and operation details

Sl. No.	Climatic conditions	Parameter	Value
1	Clear sky	Solar radiation (W/m^2)	0–950
		Ambient (°C)	24–35
		Relative humidity (%)	22–55
		Average wind velocity (m/s)	1
2	Design	Basin absorptivity (α_b)	0.90
		Basin emissivity (ε_b)	0.90
		Absorptivity of cover (α_g)	0.08
		Reflectance of cover	0.07
		Transmittance of glass (τ_g)	0.85
		Specific heat of water (C_w)	4,190 J/kg K
		Length (m)	2
		Breadth (m)	0.75
		Thickness of cotton gauze (m)	2.00×10^{-4}

Table 2. Accuracies and error for various measuring instruments

Sl. No.	Instrument	Accuracy	Range	% Error
1	Pyranometer	±30 W/m^2	0–1,800 W/m^2	3
2	Digital thermometer	±1°C	0–100°C	2
3	Thermocouple	±1°C	0–100°C	2
4	Anemometer	±0.2 m/s	0.4–30.0 m/s	1
5	Measure jar	±10 ml	0–1,000 ml	2

study was 46°C. Hot water from the reservoir enhances evaporation as the water acquires latent heat of condensation due to the regenerative effect which enhances nocturnal yield rate. The average nocturnal yield rate during the experiment was 1,571 ml/m^2. Performance of the still without cotton gauze is shown in Figure 9. Under optimal conditions, distillate yield drops to 5,600 ml/m^2. Figure 10 shows the variation in the hot water reservoir (average) temperature with respect to mass flow rate and tilt angle. The results reveal that water cooling with cotton gauze enhances the still performance.

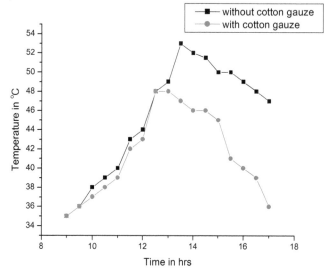

Figure 6. Variation of glass temperature.

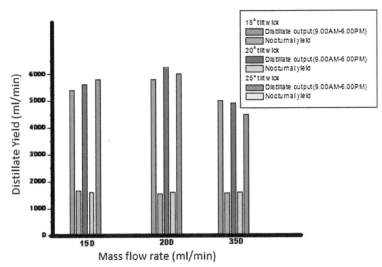

Figure 7. Variation of distillate yields with respect to mass flow rate and tilt angle (with cotton gauze).

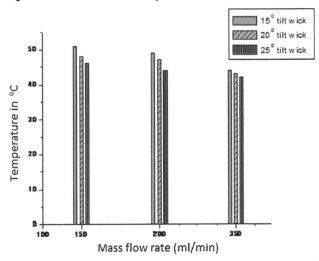

Figure 8. Variation of hot water reservoir temperature (average) with respect to mass flow rate and tilt angle (with cotton gauze).

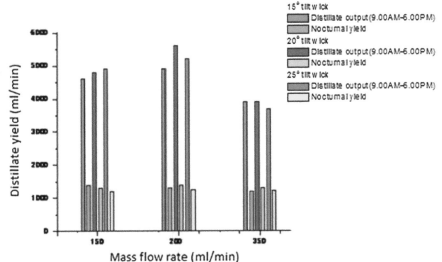

Figure 9. Variation of distillate yields with respect to mass flow rate and tilt angle (without cotton gauze).

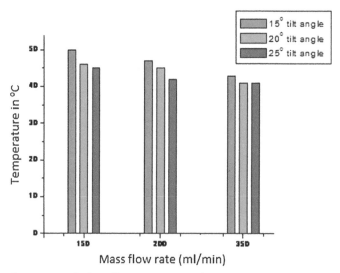

Figure 10. Variation of hot water reservoir temperature (average) with respect to mass flow rate and tilt angle (without cotton gauze).

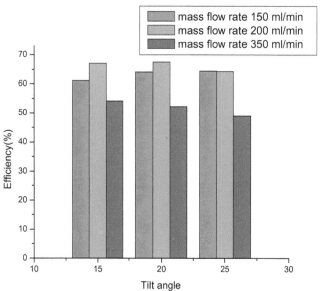

Figure 11. Variation of efficiency with respect to mass flow rate and tilt angle (with cotton gauze).

The efficiency of the still is given by:

$$\eta = \frac{ML}{IAt}$$

where M is the mass of distillate output (kg); L is the latent heat of vaporization (J/kg); I is the intensity of solar radiation (W/m²); A is the area of wick absorber (m²); t is the time (s).

Figure 11 shows the variation of efficiency with respect to mass flow rate and tilt angle for the still with cotton gauze cooling. Analysis of the graph reveals that effect of mass flow rate is more important than absorber tilt angle. The average efficiency for tilt angles 15°, 20°, and 25° with mass flow rate of 150 ml/min is 63%, the average efficiency for the same angles with mass flow rate of 200 ml/min is 66%, while the average efficiency for the same angles with mass flow rate of 350 ml/min is 52%. An increase in the 150 ml/min flow rate causes a 14% point or 22% drop in the efficiency of the still. The optimum inclination of the wick absorber was 20° for each flow rate. The wick stayed wetted regardless of the tilt angle, so different mass transfer coefficients are not likely to explain the difference.

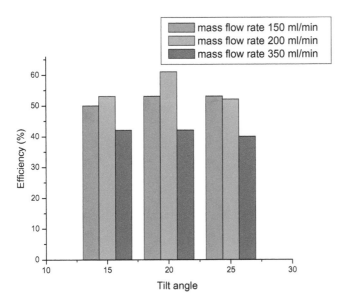

Figure 12. Variation of efficiency with respect to mass flow rate and tilt angle (without cotton gauze).

Larger tilt angles create more surface area to evaporate from, which would likely be beneficial. However, this would also indicate a lower absorber temperature, which may not be beneficial. The interaction of these effects might produce the optimum behavior. Efficiency of the still under optimum conditions of 20° tilt and 200 ml/min was 68%. The average efficiency of the still during the study was 60%. Higher mass flow rate means that the water heats up less as it moves across the absorber. This may mean lower efficiency. However, higher mass flow rate also means that the water heats up less as it moves across the glass, which would mean less loss to the air. It appears that this effect is stronger than the reduction in the efficiency at the absorber. Figure 12 shows the efficiency of the still without cotton gauze cooling. Efficiency of the still drops to 57% under optimum conditions of 20° tilt and 200 ml/min mass flow rate. The comparison of efficiencies under optimum conditions of the still with cotton gauze and without cotton gauze cooling reveals that efficiency of the still without cotton gauze cooling under regenerative effect decreases 11% points to 57%.

Comparing to a basin-type solar still of its kind (Hongfei & Xinshi, 2002) under similar conditions, we see that the efficiency is higher. This means that the advantage of regenerating the latent heat of condensation into sensible heat in the water flowing over the glass is greater than the disadvantage of losing heat to evaporation to the environment. However, it may be possible to overcome this disadvantage by having another layer of glass that would collect this evaporated water from the saline water flowing over the lower glass layer. This would have a further advantage of capturing all the water that is evaporated, meaning less supply saline water is required for a given amount of freshwater. Conversely, with the single glass layer system, if the average saline concentration is to be the same as a conventional solar still, the saline water would have to be changed more frequently. If the saline water is of limited supply and therefore cannot be changed frequently, this single glass layer system would be more appropriate for brackish water.

4. Conclusion
A "V"-type inclined wick solar still with water flowing over glass cover incorporating the regenerative effect was designed and tested under actual environmental conditions of Tamilnadu (India). The system generates fresh water, which is collected. Performance of the still due to variation in mass flow rate and angle of inclination of the wick absorber is studied. The optimum inclination of the wick absorber was 20° and mass flow rate was 200 ml/min. Efficiency of the still in optimum conditions with cotton gauze was 68% and average efficiency during the study was 60%. One main advantage of still is the improved distillate yield after sunset. In addition to the day time 9-h distillate output the system produces 1,571 ml/m² over-night

Funding
The authors received no direct funding for this research.

Author details
P. U. Suneesh[1,2]
E-mail: suneeshtvr@gmail.com
John Paul[2]
E-mail: john.me13@yahoo.com
R. Jayaprakash[3]
E-mail: jprakash_jpr@rediffmail.com
Sanjay Kumar[4]
E-mail: prof.ssinha@gmail.com
ORCID ID: http://orcid.org/0000-0001-6200-4216
David Denkenberger[5]
E-mail: david.denkenberger@gmail.com
[1] R&D Centre, Bharathiyar University, Coimbatore, Tamil Nadu,
India.
[2] Department of Science, MES College of Engineering,
Kuttippuram 679573, Kerala, India.
[3] Department of Physics, Sri Ramakrishna Mission Vidhyalaya
College of Arts and Science, Coimbatore 641020, India.
[4] Department of Physics, BR Ambedkar Bihar University,
Muzaffarpur 842001, Bihar, India.
[5] Civil and Architectural Engineering, Tennessee State
University, Knoxville TN, USA.

References
Aybar, Hikmet S., Egelioğlu, F., & Atikol, U. (2005). An
experimental study on an inclined solar water distillation
system. *Desalination, 180*, 285–289.
http://dx.doi.org/10.1016/j.desal.2005.01.009
Gorjian, S., Ghobadian, B., Tavakkoli Hashjin, T., & Banakar, A.
(2014). Experimental performance evaluation of a stand-
alone point-focus parabolic solar still. *Desalination, 352*,
1–17.
http://dx.doi.org/10.1016/j.desal.2014.08.005
Hongfei, Z., & Xinshi, G. (2002). Steady state experimental
study of a closed recycle solar still with enhanced falling
film evaporation and regeneration. *Renewable Energy, 26*,
295–308.
Janarthanan, B., Chandrasekaran, J., & Kumar, S (2006).
Performance of floating cum tilted-wick type solar still
with the effect of water flowing over the glass cover.
Desalination, 190, 51–62.
http://dx.doi.org/10.1016/j.desal.2005.08.005

Kalidasa Murugavel, K., & Srithar, K. (2011). Performance
study on basin type double slope solar still with different
wick materials and minimum mass of water. *Renewable
Energy, 36*, 612–620.
http://dx.doi.org/10.1016/j.renene.2010.08.009
Malaiyappan, P., & Elumalai, N. (2016). Productivity
enhancement of a single basin and single slope solar still
coupled with various basin materials. *Desalination and
Water Treatment, 57*, 5700–5714.
http://dx.doi.org/10.1080/19443994.2014.1003973
Minasian, A. N., & Al-Karaghouli, A. A. (1995). An improved
solar still: The wick basin type. *Energy Conversion and
Management, 36*, 213–217.
http://dx.doi.org/10.1016/0196-8904(94)00053-3
Refalo, P., Ghirlando, R., & Abela, S. (2015). The use of a solar
chimney and condensers to enhance the productivity of a
solar still. *Desalination and Water Treatment, 56*, 1–14.
http://dx.doi.org/10.1080/19443994.2015.1106096
Sakthivel, M., Shanmugasundaram, S., & Alwarsamy, T. (2010).
An experimental study on a regenerative solar still with
energy storage medium—Jute cloth. *Desalination, 264*,
24–31. http://dx.doi.org/10.1016/j.desal.2010.06.074
Sampathkumar, K., & Senthilkumar, P. (2012). Utilization
of solar water heater in a single basin solar still–An
experimental study. *Desalination, 297*, 8–19.
http://dx.doi.org/10.1016/j.desal.2012.04.012
Selvakumar, B., Kumar, S., & Jayaprakash, R. (2008).
Performance analysis of a "V" type solar still using
charcoal absorber and boosting mirror. *Desalination, 229*,
217–230.
Singh, H. N., & Tiwari, G. N. (2004). Monthly performance of
passive and active solar stills for different Indian climatic
conditions. *Desalination, 168*, 145–150.
http://dx.doi.org/10.1016/j.desal.2004.06.180
Somwanshi, A., & Tiwari, A. K. (2014). Performance enhancement
of a single basin solar still with flow of water from an air
cooler on the cover. *Desalination, 352*, 92–102.
Suneesh, P. U., Jayaprakash, R., Arunkumar, T., & Denkenberger,
D. (2014). Effect of air flow on "V" type solar still with
cotton gauze cooling. *Desalination, 337*, 1–5.
http://dx.doi.org/10.1016/j.desal.2013.12.035
Tiwari, G. N., & Bapeshwara Rao, V. S. (1984). Transient
performance of a single basin solar still with water
flowing over the glass cover. *Desalination, 49*, 231–241.
http://dx.doi.org/10.1016/0011-9164(84)85035-3
Tiwari, G. N., Madhuri., & Garg, H. P. (1985). Effect of water flow
over the glass cover of a single basin solar still with an
intermittent flow of waste hot water in the basin. *Energy
Conversion and Management, 25*, 315–322.
http://dx.doi.org/10.1016/0196-8904(85)90049-4
Velmurugan, V. J., Gopalakrishnan, M., Raghu, R., & Srithar,
K. (2000). Single basin solar still with fin for enhancing
productivity. *Energy Conversion and Management, 49*,
2602–2608.

4

Matrix cracking and delamination evolution in composite cross-ply laminates

Jean-Luc Rebière[1*]

*Corresponding author: Jean-Luc Rebière, Laboratoire d'Acoustique de l'Université du Maine, (UMR CNRS 6613), Université du Maine, Avenue Olivier Messiaen, 72085 Le Mans Cedex 9, France
E-mail: jean-luc.rebiere@univ-lemans.fr
Reviewing editor: Zude Zhou, Wuhan University of Technology, China

Abstract: This study followed numerous simulations of the stress field distribution in damaged composite cross-ply laminates, which were subjected to uni-axial loading. These results led us to elaborate an energy criterion. The related criterion, a linear fracture-based approach, was used to predict and describe the initiation of the different damage mechanisms. Transverse crack damage was generally the first observed damage. The second type of damage was longitudinal cracking and/or delamination. The stress field distribution in the damaged cross-ply laminates was analysed through an approach that used several hypotheses to simplify the damage state. The initiation of transverse cracking and delamination mechanisms was predicted. The proposed results concern the evolution of the strain energy release rate associated to the evolution of transverse cracking and delamination. As in several studies in the literature, to quantify the evolution of the damage mechanisms in the present approach, the laminate is supposed to be pre-damaged.

Subjects: Engineering & Technology, Science, Technology

Keywords: composite laminates, failure criterion, matrix cracking, delamination

1. Introduction

Composite materials are increasingly used in many structural components such as aerospace, aeronautics, automobile and sport, thanks to their height strength-to-weight ratio. It is, therefore, necessary to predict whether these structures will be able to resist under all applied stresses. A damage criterion was elaborated to evaluate the damage evolution in composite structures.

ABOUT THE AUTHOR

The activities of the laboratory are focused in most cases on "audible" acoustics, but in recent years, the laboratory also initiated research into new topics—in the field of vibrations and ultrasounds in materials. Studies concern the *spread of waves* in fluids (in repose or in flow) and in the solid (porous, granular or composite materials, vibrating structures) as well as on the *mechanisms of coupling*. Their objective is to understand physical phenomena by favouring the development of *analytical models* and of *experimental studies* linked to necessary *numerical simulation*. Researchers operate in one of three teams specialized on complementary themes: Materials; Transducers; and Vibrations, Guided Acoustics and Flow.

PUBLIC INTEREST STATEMENT

The objective of the present study is to understand the physical phenomenon of damage mode evolution in composite laminates. We propose the development of an *analytical model* with *numerical simulation*. Composite materials are used in many structural applications: aerospace, aeronautics, automobile, sport, etc. This is due to their height strength to weight ratio. It is, therefore, necessary to predict whether these structures will be able to resist all of the applied stresses. So, a damage criterion was developed to evaluate the damage evolution in composite structures. We can estimate the sequence of initiation of the different types of damage (transverse cracking, delamination and longitudinal cracking) that depend on the main parameters: material constituent, stacking sequence and loading history.

The proposed criterion is an energy-based approach. The choice of this type of criterion has been achieved after numerous studies of the stress field distribution in damaged laminates. In composite cross-ply laminates subjected to monotonic or fatigue tensile loading, the first observed damage is generally transverse cracking. This damage occurs in the central layer of the cross-ply laminate $[0_m, 90_n]_s$. This damage consists of matrix breaking between fibres in the plies perpendicular to the principal loading direction. For transverse cracking, we can experimentally observe two characteristic stages: the initiation of the damage called "first ply failure" and the limiting state when no more transverse crack can be initiated, named "characteristic damage state". The nature of the second damage mode depends on the following parameters: the laminate geometry, thicknesses of 0° or 90° layers, the nature of the fibre/matrix constituents, the loading history and the manufacturing cycle. The second damage mode, caused by high interlaminar stress levels at the interface of the layers, can be longitudinal cracking or delamination. Longitudinal cracking is similar to transverse cracking, but this damage causes matrix cracking between fibres in the layers parallel to the loading direction. Delamination is debonding between layers with different orientations. Composite structures damaged by incipient delamination or longitudinal cracking must be repaired. Experimentally, for example, in Wang and Crossman (1980), the initiation and growth of delamination were observed in a thick composite laminate. Ply separation is caused by the increase of interlaminar normal and shearing stresses. In thin composite cross-ply laminates, the damage mode succession is different. In Jamison, Schulte, Reifsnider, and Stinchcomb (1984), the second damage mode, which follows transverse cracking, is a longitudinal cracking. In this case, local delamination appears between 0° and 90° layers very late. In every case with the evolution of all the damage states, all the different damage modes cause fibre breaking in the 0° layers. All fibre breaks entail "splitting" which appears just before the ultimate failure of the composite laminate. The main objective of this work is to study the initiation and evolution of transverse crack and delamination damage.

For modelling the strain/stress relationship during damage growth, many analytical and numerical approaches have been proposed. Some models can describe the initiation of the first damage mode, transverse cracking. They mainly rely on some stress field distribution and a relationship between loading and crack density is usually proposed. The simplest model, called the "shear lag analyses", for example, (Steif, 1984), usually involves elementary assumptions regarding the displacement and stress distributions. Other models such as variational approaches use the principle of minimum complementary energy (Hashin, 1985; Vasil'ev & Duchenco, 1970). Other studies are based on finite element method (Herakovich, Aboudi, Lee, & Strauss, 1988; Rebière, 1992). Some models are based on phenomenological approaches (Allix, Ladevèze, & Le Dantec, 1990), self-consistent analyses (Adali & Makins, 1992) or approaches relying on specific aspects of the cracks (Kaw & Besterfield, 1992). As explained previously, the longitudinal cracking damage is similar to transverse cracking damage, but arises in the layers where fibres are parallel to the main loading direction, but longitudinal cracks are not always continuous (Rebière, 1992). In some laminates, longitudinal cracking does not occur before the end of life of the structure. In other laminates, longitudinal cracks can appear before the ultimate failure of the laminate. For these reasons, the investigation of longitudinal cracking is often ignored by many models in the literature. In the present work, relying on experimental observations, we suppose that the longitudinal cracks are continuous and that they span the whole length of the studied specimen.

For the study of delaminated damage, a delaminated surface with a triangular shape at the crossing of longitudinal and transverse cracks was used. The initiation of the interface debonding between orthogonal plies is estimated with the study of the evolution of the size of the triangular delamination in the x and y directions. A similar method was used by Nairn (1989).

In the literature, several approaches have been proposed to investigate the evolution of the different types of damage in composite cross-ply laminates and several kinds of criteria have been proposed (Farrokhabadi, Hosseini-Toudeshky, & Mohammadi, 2011; Rebière, Maâtallah, & Gamby, 2001), among them maximum stress-based approaches. Other kinds of criteria (Akshantala & Talreja, 2000; Barbero & Cortes, 2010; Mc Cartney, 2005; Moure, Sanchez-Saez, Barbero, & Barbero, 2014;

Nairn, 1989; Rebière & Gamby, 2004; Yokozeki, Aoki, & Ishikawa, 2005) rely on the energy release rates associated with each type of damage. Our interest in damage mechanism evolution and succession lead us to bring out the respective contributions of the transverse or longitudinal damage mechanism development which can be found in the strain energy release rate (Rebière & Gamby, 2004). The present study is restricted to damage growth in cross-ply laminates. Here, we again use the decomposition of the strain energy of the whole laminate. This analysis relies upon some estimates of the role of each strain energy component in the initiation and propagation of a given damage mechanism, such as transverse cracking, longitudinal cracking or delamiantion.

2. Model

The studied laminate is the specimen confined to a $[0_m, 90_n]_s$, composite cross-ply laminate as represented in Figure 1. The parameter used to describe the laminate architecture is the constraining parameter λ coefficient ($\lambda = t_0/t_{90}$ where t_0 is the 0° ply thickness and t_{90} is the 90° ply thickness). With the proposed approach by hypothesis, transverse and longitudinal cracks are taken continuously. The energy model used for evaluating the strain energy release rate gives good results for small stiffness laminates. However, although the proposed approach is successful for thin laminates, for thicker and more rigid laminates, the method gives approximate good results. Based on linear elastic fracture mechanics, the estimated values of the strain energy release rates are computed in a pre-damaged laminate, a method used in several damage models in the literature. Thus, there are already pre-existing transverse and longitudinal cracks. Then, the progression of transverse cracking damage is described in the following way. We consider a laminate with a periodic array of transverse cracks in the inner 90° layer. Damage initiation occurs when the spacing between two consecutive cracks is very large (*infinite*). In Rebière (1992), the strain energy release rates associated with two related problems are compared: a single transverse crack across the specimen width and two consecutive transverse cracks which span the whole width of the laminate. The equivalence of the two problems was assessed. For studying longitudinal cracking with the continuous crack hypothesis, a similar method can be used. The laminate is supposed to be "pre-cracked". The problem at hand is thus to compute the strain energy release rate for a laminate which is supposed to be previously damaged by transverse and longitudinal cracks, see Figure 1 for the whole cross-ply laminate.

The accepted assumptions for the crack geometries in the two types of layers of the laminate are as follows. The crack surfaces are supposed to have a rectangular plane geometry. Each crack extends over the whole thickness and the whole width of the 90° damaged ply. Similar assumptions are made for the longitudinal cracks in the two 0° layers. With these assumptions, it is sufficient to study only the "unit damaged cell". This "unit damaged cell" thus lies between two consecutive transverse and longitudinal cracks. In Rebière et al. (2001), the summary of the method is exposed to estimate the stress field distribution in the cracked laminate. The proposed analytical model is

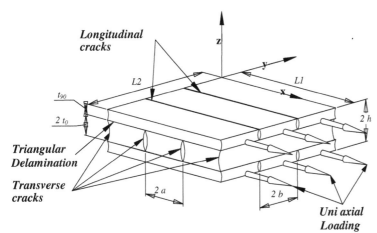

Figure 1. Laminate damaged by transverse cracks, longitudinal cracks and triangular delamination.

based on a variational approach relying on the proper choice of a statically admissible stress field (Rebière et al., 2001). In the damaged laminate, the stress field in the two layers has the following form:

$$\sigma_{ij}^{T(k)} = \sigma_{ij}^{0(k)} + \sigma_{ij}^{P(k)} \tag{1}$$

In the undamaged laminate loaded in the x direction, the layers experience a uniform plane stress state $\sigma_{ij}^{0(k)}$ obtained by the laminate plate theory (where k is the ply index, $k=0°, 90°$). The orthogonal cracks induce stress perturbations in the $0°$ and $90°$ layers which are denoted $\sigma_{ij}^{T(k)}$. In the present approach, for the sake of simplicity, thermal stresses are not taken into account.

3. Strain energy release rate

As explained previously, the laminate is supposed to be damaged by "pre-existing" transverse and longitudinal cracks. The size of the unit damaged cell depends on the transverse and longitudinal damage levels in the two types of layers, $90°$ and $0°$ layers. The strain energy release rate G associated with the initiation and development of intra-ply cracking for a given stress state is defined by the following expression:

$$G = \frac{d}{dA} \tilde{U}(\sigma, A) \quad \text{with} \quad \tilde{U}_d = N.M.U_{cel} \tag{2}$$

where $\tilde{U}_d$ is the strain energy of the whole laminate and A is the cracked area. Let L_1 denote the laminate length in the x direction and L_2 its width in the y direction. The strain energy in the damaged unit cell is denoted by U_{cel}. $N(N=L_1/2\bar{a}t_{90})$ is the number of transverse cracks and M ($M=L_2/2\bar{b}t_{90}$) is the number of longitudinal cracks. Dimensionless quantities are defined by, $\bar{a}=a/t_{90}, \bar{b}=b/t_{90}$. The crack area is $A=L_1L_2(1/\bar{a}+\lambda/\bar{b})$.

The strain energy release rates associated with transverse and longitudinal cracking are denoted G_{FT} and G_{FL}, respectively. The transverse (*resp. longitudinal*) cracking growth is characterized by the increase of the transverse (*resp. longitudinal*) crack surface initiated in the $90°$ (*resp. 0°*) layers. All details are given in Akshantala and Talreja (2000). Then:

$$G_{FT} = \frac{d\tilde{U}_d}{dA} = \frac{d\tilde{U}_d}{d\bar{a}}\frac{d\bar{a}}{dA} \quad G_{FL} = \frac{d\tilde{U}_d}{dA} = \frac{d\tilde{U}_d}{d\bar{b}}\frac{d\bar{b}}{dA} \tag{3}$$

The strain energy release rates associated with delamination is G_{del}, we get:

$$G_{del} = \frac{d\tilde{U}_d}{d\bar{d}_l}\frac{d\bar{d}_l}{dA_d} \tag{4}$$

For the analysis of the delamination evolution, only isosceles triangular geometries of the debonded area are studied. Using the present approach, the respective contribution of each damaged layer to the whole strain energy release rate can be estimated. In some models in the literature, when significant hypotheses are used to model the stress field distribution, the strain energy is only attributable to the normal stress in the damaged layer, so only the initiation of transverse cracks can be estimated with this unsophisticated type of model.

4. Results

All the numerical simulations are carried out for a prescribed uni-axial loading of 150 MPa. The T300/914 graphite/epoxy material system is studied in the following numerical computations. The proposed results show the variation of the G_{FT} and G_{del} strain energy release rates. The strain energy release rate is plotted against transverse crack density (cm^{-1}) and delaminated length for the carbon epoxy composite laminates $[0_2, 90_2]_s$. Using the above result, one can study the evolution of the strain energy release rates G_{FT} or G_{del} associated with the multiplication of transverse cracks and the evolution of the triangular delaminated surface (Table 1).

Table 1. Mechanical properties and ply thickness of T300/914 graphite epoxy system and glass epoxy system		
	Graphite epoxy system	**Glass epoxy system**
E_{LT} (GPa)	140	41.7
$E_{TT'}$ (GPa)	10	13
G_{LT} (GPa)	5.7	3.4
$G_{TT'}$ (GPa)	3.6	4.58
ν_{LT}	.31	.3
$\nu_{TT'}$	.58	.42
Ply thickness (mm)	.125	.203

In Figure 2, the results of the variation of the strain energy release rate give an idea of the influence of a pre-existing damage on the evolution of the transverse cracking and delamination damage. This result highlights the risk of the presence of a pre-existing local delamination on the transverse crack damage. At the initiation of the transverse damage, if the triangular delaminated damage is about .2 (d/a), this delaminated surface creates at the initiation of the transverse cracks a very important increase of G_{FT}. This physically corresponds to an early onset of transverse cracks in the case of laminates containing pre-existing local delamination. In the case of a large delaminated

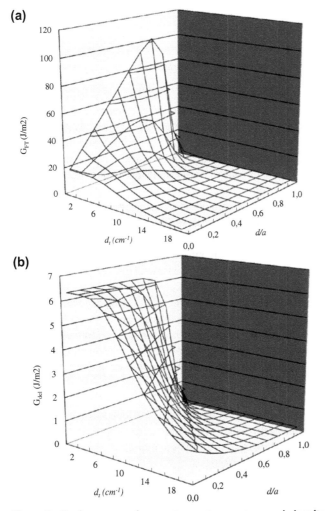

(a)

(b)

Figure 2. Strain energy release rates vs. transverse crack density and delaminated length for a graphite epoxy laminates: [0$_2$, 90$_2$]$_s$ (a) G_{FT} and (b) G_{del}.

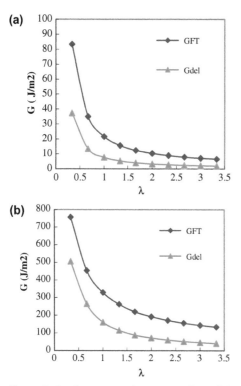

Figure 3. Strain energy release rates G_{FT} and G_{del} vs. constraining parameter λ (a) graphite/epoxy system and (b) glass/epoxy system.

area ($d/a > .7$), the strain energy release rate associated to the initiation of transverse crack tends to decrease. It is so difficult to see the initiation of transverse cracks in this case. At this stage of the delamination damage (*very large delaminated surface*), the transfer of all the stress between the two types of layers is not achieved and the central 90° layer is practically not loaded. During the multiplication of the transverse cracks, the influence of pre-existing local delaminations on the strain energy release rate G_{FT} is not always the same. From transverse crack density value of .6, the presence of local delamination causes decrease of the strain energy associated to transverse crack damage. So it becomes difficult to initiate new transverse cracks. The strain energy associated to delamination damage G_{del} decreases with the transverse crack damage growth. This allows us to say that the multiplication of transverse cracks stops the development of local delamination. This result is in the same direction with experimental results of Wang and Crossman (1980).

In Figure 3, the variation of the initiation of the strain energy release rate is exposed as a function of the constraining parameter λ. The variation of the strain energy release rate is similar for the two presented materials. For other materials, not presented in the article, the only difference lies in the numerical values of the strain energy release rates. With the present results, we can note that it will be easier to cause damage in a glass epoxy laminate than in a carbon epoxy laminate.

5. Conclusion

The curves displayed confirm that transverse cracking first occurs in the 90° layers. The results of the numerical simulations are in good agreement with experimental data and confirm that when there are incipient delaminations in a laminate resulting from the laminate manufacturing, the evolution of matrix cracking damage will quickly become very dangerous.

In this approach, the laminate is supposed to be "pre-damaged" by transverse, longitudinal cracks and delamination to investigate the initiation and development of transverse cracking and delamination. It is shown that, for properly describing the damage process, a refined computation of all the normal stresses and the inclusion of some shear stress components are necessary. Two material systems studied are proposed; for other materials, similar variations of the strain energy release rate are obtained with different numerical values.

Funding
The authors received no direct funding for this research.

Author details
Jean-Luc Rebière[1]
E-mail: jean-luc.rebiere@univ-lemans.fr
[1] Laboratoire d'Acoustique de l'Université du Maine, (UMR CNRS 6613), Université du Maine, Avenue Olivier Messiaen, 72085 Le Mans Cedex 9, France.

References
Adali, S., & Makins, R. K. (1992). Effect of transverse matrix cracks on the frequencies of unsymmetrical, cross-ply laminates. *Journal of the Franklin Institute, 329,* 655–665. http://dx.doi.org/10.1016/0016-0032(92)90078-U

Akshantala, N.-V., & Talreja, R. (2000). A micromechanics based model for predicting fatigue life of composite laminate. *Materials Science and Engineering: A, 285,* 303–313. http://dx.doi.org/10.1016/S0921-5093(00)00679-1

Allix, O., Ladevèze, P., & Le Dantec, E. (1990). Modélisation de l'endommagement d'un pli élémentaire des composites stratifiés [Damage modelization of the laminates ply]. In G. Fantozzi & P. Fleishman (Eds.), *Proceedings of 7th Journées Nationales des Composites, Lyon* (pp. 715–724). Paris: AMAC.

Barbero, E. J., & Cortes, D. H. (2010). A mechanistic model for transverse damage initiation, evolution, and stiffness reduction in laminated composites. *Composites Part B: Engineering, 41,* 124–132. doi:10.1016/j.compositesb.2009.10.001

Farrokhabadi, A., Hosseini-Toudeshky, H., & Mohammadi, B. (2011). A generalized micromechanical approach for the analysis of transverse crack and delamination in composite laminate. *Composite Structures, 93,* 443–455. http://dx.doi.org/10.1016/j.compstruct.2010.08.036

Hashin, Z. (1985). Analysis of cracked laminates: A variational approach. *Mechanics of Materials, 4,* 121–136. http://dx.doi.org/10.1016/0167-6636(85)90011-0

Herakovich, C. T., Aboudi, J., Lee, S. W., & Strauss, E. A. (1988). Damage in composite laminates: Effects of transverse cracks. *Mechanics of Materials, 7,* 91–107. http://dx.doi.org/10.1016/0167-6636(88)90008-7

Jamison, R.-D., Schulte, K., Reifsnider, K.-L., & Stinchcomb, W.-W. (1984). Characterization and analysis of damage mechanisms in tension–tension fatigue of graphite/epoxy laminates. In *Effects of defects in composites materials*, ASTM STP 836 (pp.21–55). Philadelphia, PA: American Society for Testing and Materials.

Kaw, A. K., & Besterfield, G. H. (1992). Mechanics of multiple periodic brittle matrix cracks in unidirectional fiber-reinforced composites. *International Journal of Solids and Structures, 29,* 1193–1207. http://dx.doi.org/10.1016/0020-7683(92)90231-H

Mc Cartney, L. N. (2005). Energy-based prediction of progressive ply cracking and strength of general symmetric laminates using an homogenisation method. *Composites Part A: Applied Science and Manufacturing, 36,* 119–128. http://dx.doi.org/10.1016/j.compositesa.2004.06.003

Moure, M. M., Sanchez-Saez, S., Barbero, E., & Barbero, E. J. (2014, November). Analysis of damage localization in composite laminates using a discrete damage model. *Composites Part B: Engineering, 66,* 224–232.

Nairn, J. A. (1989). The strain energy release rate of composite micro cracking: A variational approach. *Journal of Composite Materials, 23,* 1106–1129. http://dx.doi.org/10.1177/002199838902301102

Rebière, J.-L. (1992). *Modélisation du champ des contraintes créé par des fissures de fatigue dans un composite stratifié carbone/polymère* [Modeling the stress field created by fatigue cracks in a carbon/polymer composite laminate] (PhD dissertation). Université de Poitiers, France.

Rebière, J.-L., & Gamby, D. (2004). A criterion for modelling initiation and propagation of matrix cracking and delamination in cross-ply laminates. *Composites Science and Technology, 64,* 2239–2250. doi:10.1016/j.compscitech.2004.03.008

Rebière, J.-L., Maâtallah, M.-N., & Gamby, D. (2001). Initiation and growth of transverse and longitudinal cracks in composite cross-ply laminates. *Composite Structures, 53,* 173–187. http://dx.doi.org/10.1016/S0263-8223(01)00002-2

Steif, P. S. (1984). Parabolic shear-lag analyses of a [0/90]ₛ laminate. In S. L. Ogin, P. A. Smith, & P. W. R. Beaumont (Eds.), *Transverse ply crack growth and associated stiffness reduction during the fatigue of a simple crossply laminate* (Report CUED/C/MATS/TR 105, pp. 40–41). London: Cambridge University.

Vasil'ev, V. V., & Duchenco, A. A. (1970). Analysis of the tensile deformation of glass-reinforced plastics. *Mekhanica Polimerov, 1,* 144–147.

Wang, A. S. D., & Crossman, F. W. (1980). Initiation and growth of transverse cracks and edge delamination in composite laminates. Part I: An energy method. *Journal of Composite Materials, 14,* 71–87. http://dx.doi.org/10.1177/002199838001400106

Yokozeki, T., Aoki, T., & Ishikawa, T. (2005). Consecutive matrix cracking in contiguous plies of composite laminates. *International Journal of Solids and Structures, 42,* 2785–2802. http://dx.doi.org/10.1016/j.ijsolstr.2004.09.040

5

Corrosion behavior of friction stir welded AZ31B Mg alloy - Al6063 alloy joint

B. Ratna Sunil[1]* and G. Pradeep Kumar Reddy[2]

*Corresponding author: B. Ratna Sunil, Department of Mechanical Engineering, Rajiv Gandhi University of Knowledge Technologies (AP-IIIT), Nuzvid 521202, India
E-mails: bratnasunil@gmail.com, bratnasunil@rgukt.in
Reviewing editor: Duc Pham, Queen's University Belfast, UK

Abstract: In the present work, AZ31B Mg alloy and Al6063 alloy rolled sheets were successfully joined by friction stir welding. Microstructural studies revealed a sound joint with good mechanical mixing of both the alloys at the nugget zone. Corrosion performance of the joint was assessed by immersing in 3.5% NaCl solution for different intervals of time and the corrosion rate was calculated. The joint has undergone severe corrosion attack compared with both the base materials (AZ31B and Al6063 alloys). The predominant corrosion mechanism behind the high corrosion rate of the joint was found to be high galvanic corrosion. From the results, it can be suggested that the severe corrosion of dissimilar Mg-Al joints must be considered as a valid input while designing structures intended to work in corroding environment.

Subjects: Corrosion-Materials Science; Mechanical Engineering; Metals & Alloys

Keywords: FSW; Mg alloys; Al alloys; dissimilar weld joints; corrosion

1. Introduction

Recently, friction stir welding (FSW), a solid state joining technique has gained wide popularity as a promising method to join similar and dissimilar metals without melting the base materials (Bergmann, Petzoldt, Schürer, & Schneider, 2013; Mishra & Ma, 2005). Joining of light metals such as magnesium (Mg) alloys and aluminium (Al) alloys by conventional welding techniques is complex due to their high reactive nature and the difference in solidification behavior of each base material (Messler, 2014). FSW has shown its potential as a promising welding technique to join Mg and Al alloys as reported in the literature. Formation of intermetallics was observed in the weld joint of AZ31B–AA5083 (McLean, Powell, Brown, & Linton, 2003; Sato et al., 2010). Very recently, Shen, Li, Zhang, Peng, and Wang (2015) demonstrated joining of AZ31B/Al6061 alloys by FSW and reported that the strength of the dissimilar weld joints was reduced. Gerlich, Su, and North (2005) have done

ABOUT THE AUTHORS

Dr. B. Ratna Sunil is working as a faculty in the Department of Mechanical Engineering, Rajiv Gandhi University of knowledge Technologies, Nuzvid, India. His research areas include nanostructuring of metals, biomaterials, welding and processing of non-ferrous metals and surface engineering.

G. Pradeep Kumar Reddy is working as a faculty in the Department of Mechanical Engineering, Vignana Bharathi Institute of Technology, Hyderabad, India. His research areas include welding and processing of non-ferrous metals.

PUBLIC INTEREST STATEMENT

Welding of dissimilar non-ferrous metals is complex by liquid state methods. Recently, FSW, a solid state joining method has been developed to join similar and dissimilar metals. FSW is gaining wide popularity in the welding industry as a promising method to address many issues which are associated with the liquid state joining methods. The corrosion mechanisms of the weld joint are different compared with the base metals when two dissimilar metals are joined and exposed to corroding environment. In the present study, magnesium and aluminium alloy sheets were successfully joined by FSW and the corrosion behavior of the weld joint was investigated by immersing the weld joints in a corroding solution.

spot welding of AZ91 and Al6111 alloy sheets by friction stir spot welding (FSSW) and distribution of α-Mg and $Mg_{17}Al_{12}$ as eutectic mixture was noticed in the joint. AA 2024-T3/AZ31B-H24 (Cao & Jahazi, 2010), AZ31B-Al6061-T6 (Fu et al., 2015; Masoudian, Tahaei, Shakiba, Sharifianjazi, & Mohandesi, 2014) and AZ31B-Al6022 (Rao, Yuan, & Badarinarayan, 2015) are the other combinations of Mg and Al alloys which have been successfully joined by FSW. However in the literature, information on the corrosion behavior of Mg-Al dissimilar joint is insufficient. Since Mg is highly reactive metal compared with Al, understanding the corrosion behavior of Mg-Al joint is crucial particularly if the weld structure is intended to work in a corroding environment. Therefore, in the present study, AZ31B and Al6063 alloy sheets were joined by FSW and the corrosion behavior of the joint was assessed by immersion studies.

2. Experimental details

AZ31B Mg alloy sheets (2.75% Al, 0.91% Zn, 0.01% Mn and remaining being Mg) were obtained from Exclusive Magnesium, Hyderabad, India; and Al6063 alloy sheets (0.5% Mg, 0.3% Si and remaining being Al) were purchased from Metro Aluminium, Vijayawada, India. FSW was carried out by using a non consumable FSW tool made of H13 tool steel. The tool has a shoulder diameter of 15 mm and a tapered pin with 3–1 mm diameter over 3 mm length. FSW was done using an automated universal milling machine (Bharat Fritz Werner Ltd., India). The work pieces ($3 \times 50 \times 100$ mm³) were fixed on the work table of the milling machine as shown in Figure 1 and the rotating FSW tool was inserted into the joint and plunged along the joint. Initial trial experiments were conducted to optimize the process parameters and defect free joint was successfully obtained at 1,100 tool rotational speed with 15 mm/min tool travel speed. Specimens were cut at different regions of the joint using a CNC wire cut electric discharge machine (EDM, Electronica Machine Tools, India). The specimens were then metallographically polished and etched with picric acid reagent. Microstructural observations were carried out using a polarized optical microscope (Leica, Germany). Specimens of size $10 \times 10 \times 3$ mm³ were cut from both the base materials (AZ31B and Al6063 alloys) and from the centre of the weld joint. Weight of the specimens was measured by using a simple electronic balance. Corrosion studies were carried out by immersing the specimens in 3.5% NaCl solution. After 12, 24 and 36 h, the specimens were collected, dried and respective weights were measured. Experiments were conducted in triplicates ($n = 3$). The corrosion rate of the samples was calculated as per the ASTM standard G31-72 as given below (ASTM Standard G31-72, 2004).

$$Cr \ (mm/year) = k \times \Delta W / (A \times T \times D) \tag{1}$$

where $k = 8.76 \times 10^4$, T = time of exposure in hours, A = surface area of the specimen in cm², ΔW = weight loss in g, D = density in g/cm³.

Figure 1. Photograph showing FSW set up used to join AZ31B and Al6063 alloys.

3. Results and discussion

Joining of Mg-Al alloys is a complex task in welding as they exhibit different heat conducting properties. If the heat dissipation through the base materials is not uniform, then residual stresses develop hot cracks as observed in the present study at 1,400 rpm tool rotation and 25 mm/min tool travel speed. However, by optimizing the process parameters, the amount of heat generation was altered in both the base materials and sound joint was achieved. Microstructural observations show that the joint formation between AZ31B and Al6063 sheets was mainly due to the mechanical mixing or interlocking of the material at the stir zone as shown in Figure 2(b) and (c) which is similar to what reported by Venkateswaran and Reynolds (2012). Additionally, fine grain structure can be seen in the nugget zone (Figure 2(d) and (e)) which is a general characteristic of FSWed joint.

Figure 3 shows typical photographs of the specimens after different intervals of immersion in 3.5% NaCl solution. The weld joint has undergone sever corrosion compared with both the base materials. As the immersion time was increased from 12 to 24 h and further 36 h, more amount of AZ31B alloy was degraded from the joint specimen due to the corrosion and the remaining material of the

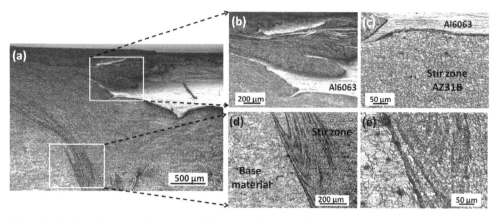

Figure 2. Optical microscope images at different regions: (a) stir zone, (b) mechanical interlocking of AZ31B and Al6063 alloys, (c) corresponding magnified image, (d) AZ31B base material and nugget zone interface and (e) corresponding magnified image.

Figure 3. Typical photographs of the specimens after immersion test: (a) Al6063 after 12 h, (b) AZ31B after 12 h, (c) weld joint after 12 h, (d) Al6063 after 24 h, (e) AZ31B after 24 h, (f) weld joint after 24 h, (g) Al6063 after 36 h, (h) AZ31B after 36 h and (i) weld joint after 36 h.

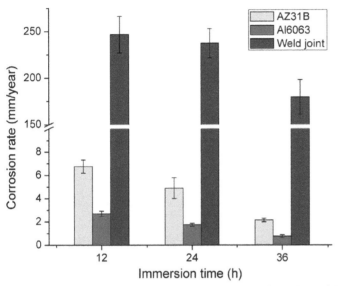

Figure 4. Corrosion rate (Cr) of the samples calculated from the weight loss after immersion test.

specimen was identified as Al6063 alloy (Figure 3(i)). The corrosion rates of the samples calculated from the weight loss measurements are shown in Figure 4. The corrosion rate of the weld joint was found to be higher compared with the corrosion rate of AZ31B and Al6063 base materials. Galvanic corrosion is believed to be the prime mechanism behind rapid corrosion of the weld joint (Donatus et al., 2015; Liu, Chen, Bhole, Cao, & Jahazi, 2009). It is true that the galvanic interactions between Mg and other metals are a serious issue in corrosion of Mg (Pardo et al., 2008) when Mg is in contact with a metal of different electrode potential. In the present study, standard electrode potential of Mg is different compared with Al (Mg −2.36 V and Al −1.66 V vs. standard hydrogen) and therefore galvanic corrosion was initiated. Also, a decreasing trend in corrosion rate was observed for all the samples as immersion time was increased from 12 to 36 h. This may be due to the oxide film formed on the sample surfaces which partially protected the substrates from the corrosion when the immersion time was increased to 36 h (Bland, King, Birbilis, & Scully, 2015; Phillips & Kish, 2013; Taheri, Phillips, Kish, & Botton, 2012). However, this oxide layer is porous and cannot provide complete protection against the corroding environment and therefore further corrosion is inevitable.

In the present study, Al alloy acted as a cathode and Mg alloy acted as an anode and in the presence of 3.5% NaCl solution, the corrosion of AZ31B was accelerated and therefore the joint was severely degraded. It has also been suggested that within the nugget zone, formation of intermetallic compounds accelerate the corrosion of α-Mg (Song & Atrens, 1999). However, in the present study, predominant reason behind higher corrosion rate of the weld joint can be claimed to the mixing of both the AZ31B and Al6063 alloys which led to raise the galvanic intensities and resulted rapid corrosion. Therefore, in developing light weight structures, joining dissimilar metals such as Mg and Al alloys may address certain issues but it is suggested based on the present study, a careful consideration of working environment is crucial. If the targeted application of the weld structure is in corroding environment, the structure may experience a sudden failure due to the rapid corrosion. Hence, from the preliminary results, it is strongly suggested to consider the galvanic corrosion effect while designing structures with weld joints of dissimilar metals.

4. Conclusions

Joining of AZ31B and Al6063 alloys was successfully done by FSW. Weld joint was found to be formed due to the mechanical interlocking of the material. Corrosion behavior of the joint was observed as higher compared with the base materials from the immersion studies. Galvanic corrosion was found to be the main mechanism behind higher corrosion rates of the weld joint. Hence from the present study, it is strongly suggested that Mg-Al joints are not suitable for applications where corrosion is suspected.

Funding
The author received no direct funding for this research.

Author details
B. Ratna Sunil[1]
E-mails: bratnasunil@gmail.com, bratnasunil@rgukt.in
G. Pradeep Kumar Reddy[2]
E-mail: pradeepmvsr04@gmail.com
[1] Department of Mechanical Engineering, Rajiv Gandhi
  University of Knowledge Technologies (AP-IIIT), Nuzvid
  521202, India.
[2] Department of Mechanical Engineering, Vignana Bharathi
  Institute of Technology, Hyderabad 501301, India.

References
ASTM Standard G31-72. (2004). *Standard practice for laboratory immersion corrosion testing of metals.* West Conshohocken: ASTM International. doi:10.1520/G0031-72R04

Bergmann, J. P., Petzoldt, F., Schürer, R., & Schneider, S. (2013). Solid-state welding of aluminum to copper—Case studies. *Welding in the World, 57,* 541–550. http://dx.doi.org/10.1007/s40194-013-0049-z

Bland, L. G., King, A. D., Birbilis, N., & Scully, J. R. (2015). Assessing the corrosion of commercially pure magnesium and commercial AZ31B by electrochemical impedance, mass-loss, hydrogen collection, and inductively coupled plasma optical emission spectroscopy solution analysis. *Corrosion, 71,* 128–145. http://dx.doi.org/10.5006/1419

Cao, X. J., & Jahazi, M. (2010). Friction stir welding of dissimilar AA 2024-T3 to AZ31B-H24 alloys. *Materials Science Forum., 638–642,* 3661–3666. http://dx.doi.org/10.4028/www.scientific.net/MSF.638-642

Donatus, U., Thompson, G. E., Zhou, X., Wang, J., Cassell, A., & Beamish, K. (2015). Corrosion susceptibility of dissimilar friction stir welds of AA5083 and AA6082 alloys. *Materials Characterization., 107,* 86–97.

Fu, B., Qin, G., Li, F., Meng, X., Zhang, J., & Wu, C. (2015). Friction stir welding process of dissimilar metals of 6061-T6 aluminum alloy to AZ31B magnesium alloy. *Journal of Materials Processing Technology, 218,* 38–47. http://dx.doi.org/10.1016/j.jmatprotec.2014.11.039

Gerlich, A., Su, P., & North, T. H. (2005). Peak temperatures and microstructures in aluminium and magnesium alloy friction stir spot welds. *Science and Technology of Welding and Joining., 10,* 647–652. http://dx.doi.org/10.1179/174329305X48383

Liu, C., Chen, D. L., Bhole, S., Cao, X., & Jahazi, M. (2009). Polishing-assisted galvanic corrosion in the dissimilar friction stir welded joint of AZ31 magnesium alloy to 2024 aluminum alloy. *Materials Characterization., 60,* 370–376. http://dx.doi.org/10.1016/j.matchar.2008.10.009

Masoudian, A., Tahaei, A., Shakiba, A., Sharifianjazi, F., & Mohandesi, J. A. (2014). Microstructure and mechanical properties of friction stir weld of dissimilar AZ31-O magnesium alloy to 6061-T6 aluminum alloy. *Transactions of Nonferrous Metals Society of China, 24,* 1317–1322. http://dx.doi.org/10.1016/S1003-6326(14)63194-0

McLean, A., Powell, G. L. F., Brown, I. H., & Linton, V. M. (2003). Friction stir welding of magnesium alloy AZ31B to aluminium alloy 5083. *Science and Technology of Welding and Joining., 8,* 462–464. http://dx.doi.org/10.1179/136217103225009134

Messler, Jr., R. W. (2014). *Principles of welding: Processes, physics, chemistry and metallurgy.* New Delhi: Wiley.

Mishra, R. S., & Ma, Z. Y. (2005). Friction stir welding and processing. *Materials Science and Engineering: R: Reports, 55,* 1–78. http://dx.doi.org/10.1016/j.mser.2005.07.001

Pardo, A., Merino, M. C., Coy, A. E., Arrabal, R., Viejo, F., & Matykina, E. (2008). Corrosion behaviour of magnesium/aluminium alloys in 3.5wt.% NaCl. *Corrosion Science., 50,* 823–834. http://dx.doi.org/10.1016/j.corsci.2007.11.005

Phillips, R. C., & Kish, J. R. (2013). Nature of surface film on matrix phase of Mg alloy formed in water. *Corrosion, 69,* 813–820. http://dx.doi.org/10.5006/0938

Rao, H. M., Yuan, W., & Badarinarayan, H. (2015). Effect of process parameters on mechanical properties of friction stir spot welded magnesium to aluminum alloys. *Materials and Design, 66,* 235–245. http://dx.doi.org/10.1016/j.matdes.2014.10.065

Sato, Y. S., Shiota, A., Kokawa, H., Okamoto, K., Yang, Q., & Kim, C. (2010). Effect of interfacial microstructure on lap shear strength of friction stir spot weld of aluminium alloy to magnesium alloy. *Science and Technology of Welding and Joining., 15,* 319–324. http://dx.doi.org/10.1179/136217109X12568132624208

Shen, J., Li, Y., Zhang, T., Peng, D., & Wang, D. (2015). Preheating friction stir spot welding of Mg/Al alloys in various lap configurations. *Science and Technology of Welding and Joining., 20,* 1–10. http://dx.doi.org/10.1179/1362171814Y.0000000248

Song, G. L., & Atrens, A. (1999). Corrosion mechanisms of magnesium alloys. *Advanced Engineering Materials., 1,* 11–33. http://dx.doi.org/10.1002/(ISSN)1527-2648

Taheri, M., Phillips, R. C., Kish, J. R., & Botton, G. A. (2012). Analysis of the surface film formed on Mg by exposure to water using a FIB cross-section and STEM-EDS. *Corrosion Science., 59,* 222–228. http://dx.doi.org/10.1016/j.corsci.2012.03.001

Venkateswaran, P., & Reynolds, A. P. (2012). Factors affecting the properties of Friction Stir Welds between aluminum and magnesium alloys. *Materials Science and Engineering A., 545,* 26–37. http://dx.doi.org/10.1016/j.msea.2012.02.069

6

Analysis and simulation of Wiseman hypocycloid engine

Priyesh Ray[1] and Sangram Redkar[1]*

*Corresponding author: Sangram Redkar, Department of Engineering, Arizona State University, Mesa, AZ 85212, USA
E-mail: sredkar@asu.edu
Reviewing editor: Duc Pham, University of Birmingham, UK

Abstract: This research studies an alternative to the slider-crank mechanism for internal combustion engines, which was proposed by the Wiseman Technologies Inc. Their design involved replacing the crankshaft with a hypocycloid gear assembly. The unique hypocycloid gear arrangement allowed the piston and connecting rod to move in a straight line creating a perfect sinusoidal motion, without any side loads. In this work, the Wiseman hypocycloid engine was modeled in a commercial engine simulation software and compared to slider-crank engine of the same size. The engine's performance was studied, while operating on diesel, ethanol, and gasoline fuel. Furthermore, a scaling analysis on the Wiseman engine prototypes was carried out to understand how the performance of the engine is affected by increasing the output power and cylinder displacement. It was found that the existing 30cc Wiseman engine produced about 7% less power at peak speeds than the slider-crank engine of the same size. These results were concurrent with the dynamometer tests performed in the past. It also produced lower torque and was about 6% less fuel efficient than the slider-crank engine. The four-stroke diesel variant of the same Wiseman engine performed better than the two-stroke gasoline version. The Wiseman engine with a contra piston (that allowed to vary the compression ratio) showed poor fuel efficiency but produced higher torque when operating on E85 fuel. It also produced about 1.4% more power than while running on gasoline. While analyzing effects of the engine size on the Wiseman hypocycloid engine prototypes, it was found that the engines performed better in terms of power, torque, fuel efficiency, and cylinder brake mean effective pressure as the displacement increased. The 30 horsepower (HP) conceptual Wiseman prototype, while operating on E85, produced the most optimum results in all aspects, and the diesel test for the same engine proved to be the most fuel efficient.

Subjects: Energy & Fuels; Mechanical Engineering; Mechanical Engineering Design

Keywords: hypocycloid engine; multi-fuel analysis

ABOUT THE AUTHORS

Priyesh Ray graduated with Master of Science in Technology from Arizona State University in 2014. He completed his Bachelor's from Arizona State University in 2012. His research interests are engine design and testing.

Sangram Redkar is an associate professor in Polytechnic School at Arizona State University. He completed his PhD from Auburn University in 2005. His research interests are robotics, dynamics and control, and machine design.

PUBLIC INTEREST STATEMENT

This research investigates a performance, multi-fuel operation and scalability of a hypocycloid engine. This engine is based on Wiseman hypocycloid mechanism. In this engine, connecting rod motion is purely translational that minimizes the size loads and corresponding frictional losses observed in a typical slider-crank engine. The advantages and disadvantages of this hypocycloid engine along with simulation and experimental data are presented in this paper.

1. Introduction

Internal combustion engines (ICEs) have been around for centuries and the slider-crank mechanism is typically used to convert the reciprocating motion of piston to rotary motion of crankshaft. Despite its popularity, the slider-crank mechanism still has a few flaws, for example the higher side load and associated frictional losses in the piston cylinder assembly. Wiseman Technologies Inc (WTI) proposed an alternate mechanism as a solution to tackle this problem in the form of Wiseman hypocycloid engine (WHE). It consists of a unique gear assembly based on the hypocycloid concept.

As shown in Figure 1, a hypocycloid is a curve produced by tracing the fixed point "P" on the circumference of the small circle (of radius r_a), as it rolls inside the large circle (of radius r_b). The path of that curve is given by the following equations;

$$x = (r_a - r_b) \cos \emptyset + r_b \cos \left(\frac{r_a - r_b}{r_b} \emptyset \right) \qquad (1)$$

$$y = (r_a - r_b) \sin \emptyset - r_b \sin \left(\frac{r_a - r_b}{r_b} \emptyset \right) \qquad (2)$$

where r_a is the radius of the smaller circle; r_b is the radius of the larger circle; $\emptyset$ is the angle from the x axis to the line that intersects the center of circle "a" and circle "b".

Now, in a special case, where the ratio of the radii of the two circles is 2:1, the hypocycloid curve at any given point on the circumference of circle "b" is a straight line. When the circles are replaced by gears, whose pitch diameters have a ratio of 2:1, then such an assembly can be used to produce a perfect straight-line motion of a piston in an ICE. In Figure 2, the smaller pinion gear in red can be compared to the small circle "a" and the internal ring gear (blue) to the larger circle "b" in Figure 1. The visualization shows that a specific point on the pitch diameter of the pinion gear is always coincident with the vertical black line, as it rolls inside the ring gear. The black line indicates the path of the piston and the connecting rod.

Historically, the hypocycloid engine variants have received some attention from the scientific community. Many patents have been filed around the hypocycloid mechanism and its application to pump, compressors, and engines. In 1802, Murray used the hypocycloid mechanism to build a steam engine (Karhula, 2008). In 1975, Ishida published an excellent paper that focused on the inertial shaking forces and moments of a hypocycloid engine and compared to a conventional slider-crank

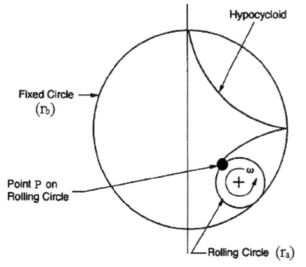

Figure 1. Hypocycloid concept (*Special plane curves*, 2005–2014).

Figure 2. Geared hypocycloid concept rotating at 45° increments (Conner, 2011).

engine (Ishida & Matsuda, 1975). Building upon Ishida's research, Drs. Beachley and Lenz published a paper that proposed an interesting method to attain perfect balance in hypocycloid engines (Beachley & Lenz, 1988).

Another important contribution to the study of hypocycloid engine came from Drs. Ruch, Fronczak, and Beachley (1991). They studied a variant of hypocycloid engine called the modified hypocycloid engine. This engine had a third gear that reduced the overall individual gear tooth loads. In 1998, Andriano designed and tested of a full crank 125cc two-stroke hypocycloid engine (Andriano, 1998). Their design had a seal around the connecting rod to isolate the crankcase from the top end. In 2008, Karhula published a paper that compared slider-crank and hypocycloid engines of equal dimensions (stroke, bore, etc.) (Karhula, 2008). For more details and a detailed literature review, the reader is referred to the thesis by Conner (2011).

Though a simple hypocycloid mechanism such as Cardan Gear produces a straight line with a perfect sinusoidal motion at any given point, the angle of the straight line depends on the point selected on the pitch of the small pinion gear. WTI modified this gear assembly and incorporated it in their hypocycloid engine, which was patented in 2001. The first WHE was built by adding a link to support the pinion gear where the rotating motion of the gear is transferred to a rotating output shaft (L2) as shown in Figure 3. The point D1 as seen in Figure 3 was connected to the bottom end of the connecting rod. This point represented the point "P" shown in Figure 1.

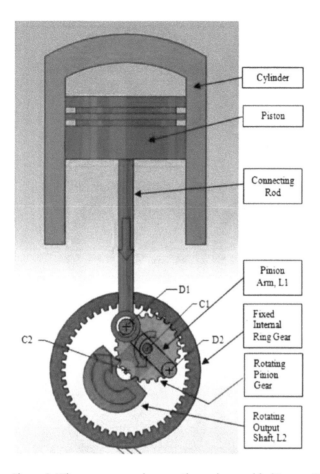

Figure 3. Wiseman gear and connecting rod assembly (Conner, 2011).

The proposed link in the Wiseman mechanism (Wiseman, 2001) to support the pinion gear (Item 200), known as the carrier shaft (Item 100), can be seen in Figure 4. The Wiseman mechanism also has a provision for the pinion gear teeth (Item 204) to mesh with the fixed internal ring gear (Item 6) in the form of a cavity (Item 322). This unique mechanical assembly allowed the piston in the Wiseman engine to travel in a perfect straight line sinusoidal motion eliminating any piston side load (Wiseman Engine Group, 2010). It also eliminated connecting rod bending and reduced the engine heat loss due to side load friction.

The conceptual Wiseman Engine is shown in Figure 5(a). As discussed earlier, the path of the crankpin in the Wiseman engine is hypocycloidal instead of circular. This provides the advantages such as longer power stroke, straight line motion better performance, and less wear.

2. Software simulations

To analyze the performance of the WHE, the Lotus Engine Simulation (LES) software was extensively used. The LES software is capable of simulating the performance of an ICE and can predict various output parameters of engine under different operating conditions. These parameters can be fuel efficiency, power output, torque, and running temperature of the engine over a period of time. Before trying a completely new hypocycloid engine concept in LES, the authors decided to validate the LES simulation results for a conventional slider-crank four-stroke (GUNT 1B30) diesel engine that was available in our laboratory. The GUNT engine is a four-stroke, single cylinder, air-cooled engine, with direct injection that works on both diesel and biodiesel fuels. The compression ratio for this engine is 21.5 with a mean piston speed of 6.9 m/s and a mechanical efficiency of 87%. For GUNT 1B30 engine, a model was created in LES with specifications as shown in Table 1. The software simulations were

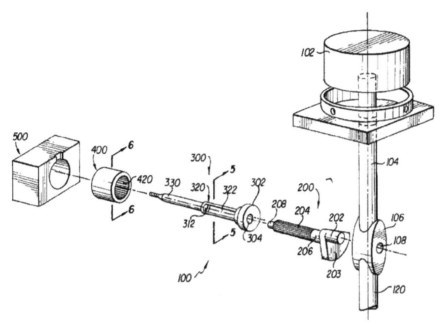

Figure 4. Patented Wiseman design (Wiseman, 2001).

Figure 5(a). Wiseman engine conceptual model (Conner, 2011).

Table 1. GUNT 1B30 engine specifications (Hatz Diesel, 2012)			
Power	Stroke	Bore Ø	Displacement
5.5 KW @ 3,500 rpm	69 mm	80 mm	347cc

carried out to obtain performance results. These software results were compared to manufacturer supplied performance data. Furthermore, to validate the authenticity of the results, the simulation results were compared with dynamometer test data.

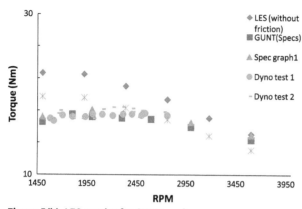

Figure 5(b). LES results for GUNT engine.

Figure 5(b) shows the LES simulation results (with and without the frictional/mechanical losses), manufacturer supplied performance plots, and experimental (dynamometer testing) results. The GUNT engine has the mechanical efficiency of 87%. LES (without friction) assumed mechanical efficiency as 100%. These LES (without friction) results were then multiplied later with 0.87 to compensate for the mechanical efficiency and called LES (with friction). It can be noticed in Figure 5(b) that the torque results recorded during the dynamometer tests coincided very well with the manufacturer specifications at 2,500–3,000 rpm. The slight variation in simulated and actual results can be observed because a perfect displacement scavenging model was selected for the simulation, which assumes that all the residual gases are removed, and only a fresh charge of air is present during the combustion. But otherwise, the LES software results are comparable to the test and manufacturer data, and show the same trend.

2.1. Modeling the Wiseman engine
Once the reliability of the LES software was established for a conventional slider-crank engine, it was used to model the WHE 30cc prototype. The Wiseman engine prototype is 30cc, two-stroke, single cylinder, with spark ignited and carbureted fuel system. It is designed to generate 0.99 HP of power at around 7,000 rpm. To model the engine in LES software, an intake disk valve was also used in addition to variable volume inlet plenum. Wiseman engine has piston ported intake and exhaust valves with dimensions shown in Table 2.

This Wiseman engine prototype was built from a 30cc weed-wacker slider-crank engine, where the slider-crank mechanism was replaced with Wiseman hypocycloid mechanism. Table 3 provides a summary of their respective design specifications.

Table 2. Wiseman LES port data	Intake	Exhaust
Port width (mm)	40	20.82
Maximum port height (mm)	2	7.41
Valve open (°)	124	108

Table 3. Summary of Wiseman and stock engine cylinder dimensions			
Engine	Stroke	Bore Ø	Displacement
Wiseman engine	1.125 in (28.6 mm)	1.435 in (36.5 mm)	1.819 in³ (29.81cc)
Stock homelite	1.114 in (28.3 mm)	1.435 in (36.5 mm)	1.802 in³ (29.53cc)

The intake and exhaust pipe geometry in the software model was assumed to be simple tubes of approximate dimensions of the carburetor nozzle, at wide-open position. A single Wiebe combustion model was used for combustion analysis. This model assumes that the heat released during the combustion heats the whole combustion chamber.

The single Wiebe function defines the mass fraction burned as:

$$1 - \exp^{-A\left(\frac{\theta}{\theta_b}\right)^{M+1}} \tag{3}$$

where the Wiebe coefficients A and M for gasoline are 10.0 and 2.0, respectively.

A simpler, Annand heat transfer model (Annand, 1963) was used for the Wiseman engine which was given by:

$$\frac{hD}{k} = ARe^B \tag{4}$$

where h is the heat transfer coefficient; A and B are the Annand open or close cycle coefficients; K is the thermal conductivity of gas in the cylinder; D is the cylinder bore diameter; Re is the Reynolds number based on the means piston speed and engine bore (The A and B coefficients for a carbureted or a port injected combustion system are 0.2 and 0.8, respectively.).

One of the key characteristics of the Wiseman engine is its hypocycloid mechanism. The LES software assumes that all the engines models designed in it follow the piston motion of a slider-crank engine, and so the results are simulated accordingly. To simulate the Wiseman engine according to the hypocycloid piston motion, a user specific subroutine was created. The equations in the subroutine were based on the piston motion equations of the Wiseman hypocycloid gear train, which provided the instantaneous position and the cylinder volume, while the piston is in the motion. These equations were derived from Figure 6.

The distance P or piston position is relative to the Top Dead Center (TDC). After finding the piston postion, the volume is found by using:

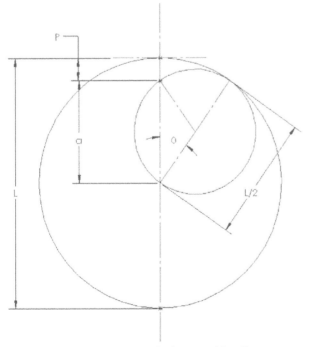

Figure 6. Wiseman (hypocycloid) piston position diagram.

$$P = \frac{L}{2} - \left(\frac{L}{2} \cos \theta \right) \tag{5}$$

$$V' = P \left(\frac{\pi B^2}{4} \right) \tag{6}$$

where L is the stroke of the engine; θ is the crank angle with 0° being TDC; V' is the volume above piston, TDC being 0; B is the diameter of the cylinder (bore); P is the position of the piston with the origin being TDC.

It was also known that the the volume of the Wiseman engine at each crank angle is lower than that of the original stock engine since its piston sits higher than the stock engine. This provides less combustion volume at each crank angle. Previous dynamometer tests on the Wiseman engine suggested that the engine operates at a mechanical efficiency of 60.6%, and this was taken into consideration while modeling the engine friction. Furthermore, a virtual sensor was attached to the cylinder block in the LES model to record the crank angle, piston speed, and position, as seen in Figure 7.

To verify the modification in the piston motion, the model was simulated using subroutines for a conventional slider-crank and the WHE. The data collected from the sensor for both the runs was recorded and compared to see the change in the piston movement, which can be seen in Figures 8 and 9.

From Figure 8, it was noticed that the piston velocity for a slider-crank engine does not trace a perfect cosine–time curve, whereas the hypocycloid piston motion syncs perfectly with the cosine curve. There is nonlinearity in the slider-crank mechanism, and so the piston motion is not a harmonic function. Similar observation can be made from Figure 9, the displacement of hypocycloid engine is harmonic but the displacement of the slider-crank engine is not.

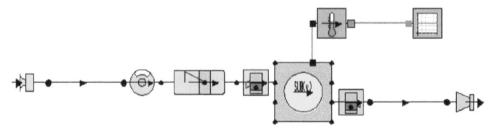

Figure 7. LES model of Wiseman engine.

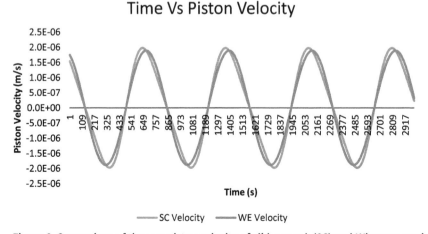

Figure 8. Comparison of time vs. piston velocity of slider-crank (SC) and Wiseman engine (WE).

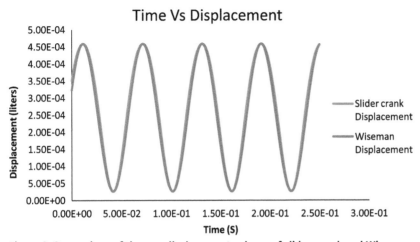

Figure 9. Comparison of time vs. displacement volume of slider-crank and Wiseman engine.

2.2. Comparison of simulation with experiments

As mentioned earlier, a weed-wacker engine was modified to incorporate Wiseman hypocycloid mechanism. This prototype engine was termed WHE alpha prototype. It is a 30cc two-stroke, single cylinder engine that is designed to produce 1 HP. This engine was modeled in LES software, and simulated for speeds ranging from 1,000 to 8,000 rpm with an increment of 1,000 rpm. A similar commercially available slider-crank engine was also simulated in LES software under identical conditions. The results from the simulations are shown in Figures 10–12.

As seen in Figure 10, the Wiseman engine at 6,000 rpm produced slightly less power than the slider-crank engine. Its power output at 6,000 rpm was 0.62 KW (0.83 HP), and the same for the slider-crank engine was 0.66 KW (0.89 HP). Similar trend can be observed as the engine RPM increased. At all speeds, slider-crank engine of 30cc capacity produced more power compared to the Wiseman engine.

Furthermore, comparing the software results for output torque of both the engines, it can be seen from Figure 11 that the torque at peak RPM of Wiseman engine is slightly less than that of the slider-crank at the same RPM. This could be because the Wiseman engine has a shorter stroke length than the slider-crank engine. The Wiseman engine at 6,000 rpm produces 0.98 Nm of torque, and the slider-crank engine produces 1.05 Nm. Similar trend can be observed as the engine RPM increased. At all speeds, slider-crank engine of 30cc capacity produced more torque compared to the Wiseman engine.

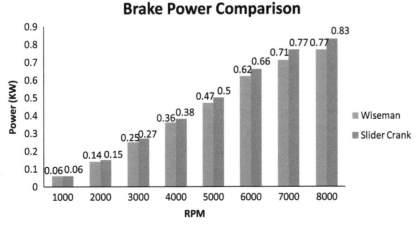

Figure 10. LES results of Wiseman and slider-crank engine for brake power.

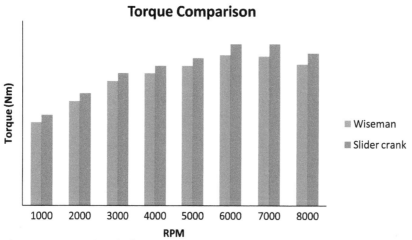

Figure 11. LES results of Wiseman and slider-crank engine for torque.

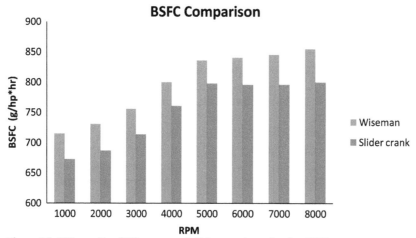

Figure 12. LES results of Wiseman and slider-crank engine for BSFC.

As shown in Figure 12, it can be noted that the Wiseman engine at 6,000 rpm resulted in a break specific fuel consumption (BSFC) of 840.42 g/hp h, while the slider-crank engine returned to 795.5 g/hp h. This shows that the Wiseman engine is about 6% less fuel efficient than the slider-crank engine. It can be observed that all speeds slider-crank engine of 30cc capacity has less BSFC compared to Wiseman engine.

The WHE alpha prototype was tested using a dynamometer at MTD Southwest Inc. in Chandler, Arizona. During the dynamometer tests, the WHE alpha prototype engine's performance was also compared to that of a commercial slider-crank (MTD) engine of the same size. A brief summary of these results can be seen in Table 4.

In the dynamometer tests, it was observed that at a peak speed of 7,000 rpm, the Wiseman engine had about 5% higher loss in power than a stock engine of the same size. The Wiseman engine was also found to be 21% less fuel efficient compared to a slider-crank engine at 6,000 rpm. The trend of higher BSFC in the Wiseman engine is concurrent with the results from the dynamometer tests.

Table 4. Wiseman dynamometer test summary (Conner, 2011)		
Engine	**Peak power**	**BSFC**
MTD engine (31cc)	0.96 HP @ 7,000 rpm	410.08 g/hp h
Wiseman engine	0.60 HP @ 6,000 rpm	520.06 g/hp h

3. Multi-fuel and scaling of WHE

To explore the range of applications of the Wiseman engine, it was necessary to analyze its performance while operating on different fuels. For the same purpose, a four-stroke diesel variant of the Wiseman engine was modeled and simulated in LES. To simulate a four-stroke diesel Wiseman engine, the existing LES software model for the 30cc two-stroke engine was modified. The physical parameters of the engine such as the bore and stroke dimensions, swept volume, and connecting rod length were kept unchanged, but the compression ratio was increased from 8:1 to 17:1. Also, the fuel delivery system was changed to direct injection type instead of carbureted. The engine results were then simulated under the similar test conditions as the original gasoline engine. The results for engine's output power, torque, and fuel consumption can be seen in the following plots.

From the data in Figures 13–15, it can be noticed that the four-stroke diesel Wiseman engine performed better than both the two-stroke Wiseman engine (using gasoline) and the two-stroke slider-crank engine. This improvement in the performance was consistent for power (Figure 13), torque (Figure 14), and fuel consumption (BSFC) at peak RPMs (Figure 15). The improvement can be credited to the higher volumetric efficiency of the four-stroke engine, which reduced the loss of air–fuel mixture during the combustion resulting in better fuel efficiency (Ganesan, 2012, p. 638). The proper utilization of the air in the four-stroke engine also resulted in increased power output. Furthermore, the absence of piston ports increased the effective cylinder fuel compression range in the four-stroke engine (Table 5).

WTI (n.d.) has also proposed a VCR-Variable Compression engine to be used in an Unmanned Arial Vehicle (UAV). The engine's design incorporated an adjustable contra piston which enables the operator to change the engine's compression ratio while it is running. This engine has a modified

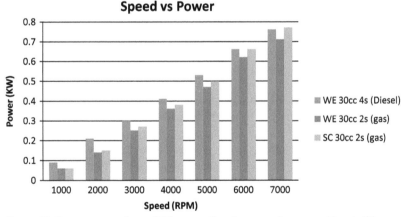

Figure 13. Power comparison of Wiseman diesel, gas, and a conventional slider-crank gas engines.

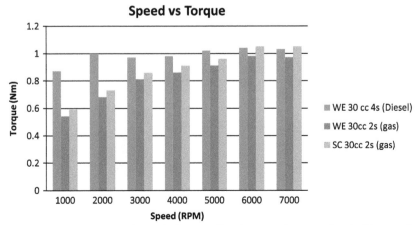

Figure 14. Torque comparison of Wiseman diesel, gas, and conventional slider-crank gas engines.

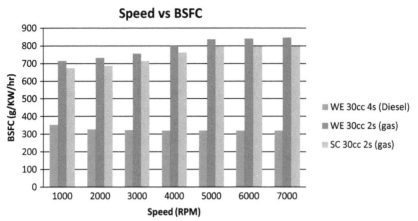

Figure 15. Fuel efficiency comparison of Wiseman diesel, gas, and conventional gas engines.

Table 5. Performance summary at the peak speed of 7,000 rpm			
Engine	Power (KW)	Torque (Nm)	BSFC (g/KW/h)
Wiseman two-stroke (gas)	0.71	0.97	845.08
Wiseman four-stroke (diesel)	0.76	1.03	319.62
Slider-crank two-stroke (gas)	0.77	1.05	796.2

cylinder head with contra piston acting as the bottom face of the combustion chamber. By adjusting the distance travelled by the contra piston with the help of a bolt, the effective cylinder clearance volume can be changed. This results in change of the compression ratio. The ability to change the compression ratio of the engine allows it to operate on variety of fuels without knocking. The modified cylinder head and the contra piston design can be seen in Figure 16 below.

An approach similar to the diesel engine model was adopted to simulate the VCR Wiseman UAV engine. Since the contra piston design only changes the compression ratio of the engine, the software model for the two-stroke Wiseman engine (alpha prototype) was kept unchanged except for the compression ratio. A simulation was carried out to analyze the performance of the Wiseman engine while operating on Ethanol (E85) fuel. E85 is a blend of gasoline and ethanol, with 85% ethanol and 15% gasoline by volume. It is also known as flex-fuel (US Department of Energy, Energy Efficiency & Renewable Energy, 2013). It has a higher octane rating than gasoline and is believed to produce more power. Studies have shown an increase of about 5% output power by adding 10%

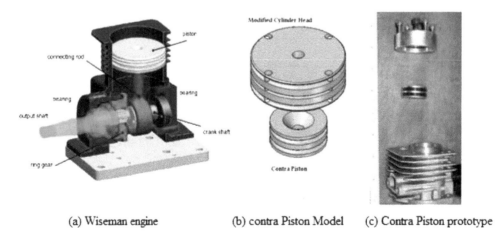

(a) Wiseman engine   (b) contra Piston Model   (c) Contra Piston prototype

Figure 16. Variable compression Wiseman UAV engine.

ethanol blend in a spark ignited combustion engine (Datta, Chowdhuri, & Mandal, 2012). Increasing the concentration of the ethanol in the fuel blend also tends to increase the output power and the overall efficiency (Celik, 2007). The fuel properties of E85 considered for the simulation can be seen in Table 6 below.

Studies have shown that spark-ignited engines, while operating on ethanol blends, performed better at a compression ratio of about 12:1 to 13:1 (Costa & Sodre, 2010a). So the compression ratio in the software model for the Wiseman engine was changed to 12:1 while simulating its performance on E85 fuel. The engines utilizing ethanol have a higher BSFC, since ethanol has a lower calorific value than gasoline. This means that the engine consumes more fuel to generate the same amount of power and torque while running on ethanol, as compared to gasoline (Costa & Sodre, 2010b).

From comparing the results of the engine operating on gasoline and E85 in Figure 17, it can be noticed that the Wiseman engine produced less or identical power at lower speeds. But at the peak speed of 7,000 rpm, the output power of the engine was 1.4% more in the case of E85 as opposed to gasoline. This is typical of a port-inject gasoline engine operating on ethanol (Cahyono & Bakar, 2010). Another reason for higher power output while using E85 in the Wiseman UAV engine is because, at higher temperatures, ethanol has a better thermal efficiency than gasoline as ethanol has a higher heat of vaporization. When the compression ratio is increased, the engine burns a richer mixture of air–fuel in the case of E85, resulting in a higher output power (Costa & Sodre, 2010a).

From Figure 18, it can be noticed that the torque generated by the Wiseman UAV engine while operating on E85 is slightly higher. As mentioned earlier, this is because it had richer air–fuel which causes the fuel to burn closer to stoichiometric, resulting in a better combustion (Topgül, Yücesu, Çinar, & Koca, 2006). The simulated results for WHE running on E85 showed that an increase in

Table 6. E85 fuel properties (US Department of Energy, Energy Efficiency & Renewable Energy, 2013)	
Fuel properties of E85	
Fuel system	Direct injected
Fuel type	Ethanol
Calorific value (kJ/kg)	2,8765
Density (kg/l)	0.782
H/C ratio fuel (molar)	2.7177
O/C ratio fuel (molar)	0.3951
Molecular mass (kg/k mol)	46
Maldistribution factor	1.000

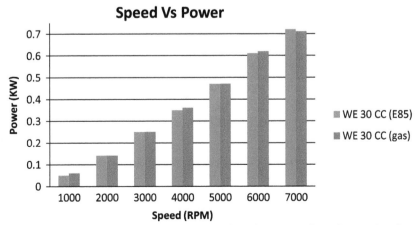

Figure 17. Power comparison of Wiseman engine with contra piston (E85 and gas).

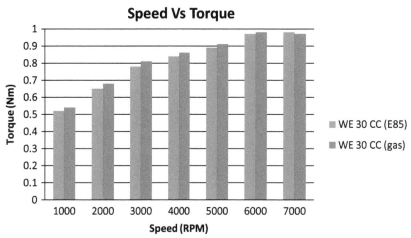

Figure 18. Torque comparison of Wiseman engine with contra piston (E85 and gas).

torque is only at the peak RPM, because at lower RPMs, the high calorific value of gasoline results in higher torque. As the engine speed increased, the ethanol blend tends to produce more torque due to its faster flame velocity. The higher compression ratio used for E85 fuel also increased the cylinder brake mean effective pressure (BMEP), so more work was done on the piston causing an increase in the output torque (Costa & Sodre, 2010b).

Studies have shown that increasing the concentration of ethanol by 10–20% in an ethanol–gasoline blend reduced the calorific value of the fuel, which caused an increase in fuel consumption (Cahyono & Bakar, 2010). In other words, more fuel is required for an engine operating on E85 blend to do the same amount of work as an engine running on pure gasoline. This trend in fuel efficiency is also evident in the results generated by the LES software. The comparison of the results in Figure 19 shows that the Wiseman UAV engine, when running on E85, tends to consume more fuel at every speed with an increase of almost 41.5% at the peak RPM (Table 7).

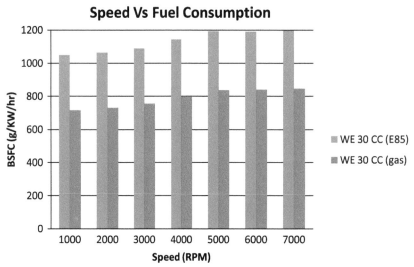

Figure 19. Fuel consumption comparison of Wiseman engine with contra piston (E85 and gas).

Table 7. Performance summary of Wiseman with contra piston (E85 and gas) at 7,000 rpm			
Engine	Power (KW)	Torque (Nm)	BSFC (g/KW/hr)
Wiseman UAV (gas)	0.71	0.97	845.08
Wiseman UAV (E85)	0.72	0.98	1,196.21

3.1. Scaling the Wiseman engine prototypes with respect to engine size

There is a need to increase the engine size of the Wiseman engine in order to expand its range of applications. The current 30cc Wiseman alpha prototype engine was designed to produce 0.99 HP, whereas engines required for electric generators, lawn mowers, UAVs, and small tractors need to produce power output much higher than 1 HP. For this purpose, a series of theoretical Wiseman prototypes were designed and their performance was simulated in LES software. These engines were also simulated using conventional slider-crank for comparison. The new theoretical prototypes were designed to produce 10, 20, and 30 HP (7.46, 14.91, and 22.37 KW, respectively). Furthermore, analysis was conducted to determine how the engine's performance varied with respect to its size, and while operating on different fuels (gasoline, diesel, and ethanol).

The engines were designed to have the following characteristics:

- Single cylinder, with a four-stroke combustion cycle, and a peak performance speed of 2,000 rpm.
- Compression ratio of 8:1 was chosen for the gasoline engines, 16:1 for diesel engines, and 13:1 for ethanol engines.
- Indicated Mean Effective Pressure (IMEP) of 0.7 MPa.
- Mechanical efficiency (η_{mech}) of 80%.
- BMEP of 0.56 MPa (IMEP $\times \eta_{mech}$).
- Bore/stroke ratio (L/D) of 1.2 for gasoline and ethanol engines and 1.25 for diesel engines.
- The gasoline and ethanol engines used a carburetor and the diesel engines were direct injected.

The following steps further explain the calculations carried out during the designing process of each prototype:

(i) Taking the above-mentioned specifications into consideration, the engine bore diameter was calculated by:

$$Bp = BMEP \times L \times A \times N \tag{7}$$

where Bp is the Brake horsepower in Watts; BMEP is the brake mean effective pressure; L is the stroke length (1.2 × D); A is the area of the bore $\left(\frac{\pi}{4}D^2\right)$; N is the engine speed (Max. RPM/2).

This step was carried out to calculate the bore diameter for each engine, and their respective stroke lengths were found.

(ii) Once the bore diameter and the stroke length was found, the engine swept volume was calculated by using:

$$Vs = A \times L \tag{8}$$

where Vs is the cylinder swept volume; A is the area of the bore $\left(\frac{\pi}{4}D^2\right)$; L is the stroke length.

(iii) Finally, torque at peak performance RPM for each engine was calculated;

$$T = \frac{HP \times 5252}{Speed} \tag{9}$$

where the horsepower (HP) is in Kilowatts (KW) and the speed at the peak performance RPM of 2,000.

3.2. Scaling laws for ICEs

Now, to analyze the performance of these prototypes, and predict how their output changes with respect to the size, some performance scaling laws were established. Since every engine's performance parameters vary over its operating range, comparing them at randomly selected points would not draw any meaningful conclusions. In order to compare the performance metrics across different engines, the engines needed to be compared at constant speed (2,000 rpm) and air–fuel ratio. Using this method, the performance can be isolated and treated as a result of change in engine size. The scaling laws were based on the laws proposed by Dr. Shyam Kumar Menon over the years of his research in engine scaling. More than 40 engines, single cylinder to 36 cylinders, and displacement ranging from 0.1cc to almost 100,000cc, were studied and tested by Dr. Menon to establish trends in the engine output based on their displacement. The following scaling laws were proposed by Menon (2010):

I. Scaling the bore and stroke of the engine

The engines with displacement less than 1,000cc are likely to have a "square" design with a few exceptions between the ranges of 1,000–8,000cc as shown in Figure 20. This meant that the bore-to-stroke ratio for those engines is close to 1. But as the engines got closer to a displacement of 1,000cc, the design changed to slightly "under squared" (Menon, 2010).

The calculated displacements of the Wiseman prototypes and its corresponding bore-to-stroke ratios in Tables 8 and 9 were plotted to verify the established scaling law. It can be seen from Figure 20 that the engines tend to have a bore-to-stroke ratio closer to 1, regardless of their displacement.

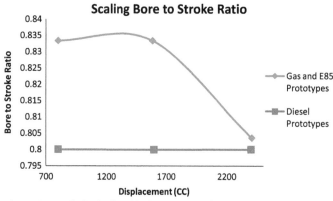

Figure 20. Scaled relationship between engine sizes and bore-to-stroke ratio.

Table 8. Summary calculated specifications for 10, 20, and 30 HP gasoline engines

Engine power (KW)	Bore diameter (mm)	Stroke length (mm)	Swept volume (l)	Calculated torque (Nm)	Bore-to-stroke ratio	Swept volume (cc)
7.46	94.7	113.64	0.800	35.6	0.83333	800
14.914	119	142.8	1.587	71.21	0.83333	1,587
22.371	135	168	2.404	106	0.80357	2,404

Table 9. Summary of calculated specifications for 10, 20, and 30 HP diesel engines

Engine power (KW)	Bore diameter (mm)	Stroke length (mm)	Swept volume (l)	Calculated torque (Nm)	Bore-to-Stroke ratio	Swept volume (cc)
7.46	93.4	116.75	0.7995	35.6	0.8	799.5
14.914	117.65	147.063	1.598	71.21	0.7999	1,598
22.371	134.68	168.35	2.397	106	0.8	2,397

Notes: An approach similar to designing the Wiseman UAV engine with contra piston was taken for ethanol engines. This meant that only the compression ratio (13:1) and fuel properties (E85) were changed in the LES models of gasoline engines.

This is especially true for the diesel engine prototypes. One reason behind this could be that the data-set used to test this particular scaling law was too small to draw any meaningful conclusion. Though, the Wiseman diesel prototypes showed a strong relationship with the Kohler diesel engines and the Ford automobile engines tested between the ranges of 700–3,000cc (approximately) (Menon, 2010).

II. Scaling the engine peak torque at peak power output

It was established that the engine torque at peak power increases with the increase in engine displacement (Menon, 2010). The output torque results generated by software simulations at peak power for each engine prototype was plotted to verify this relationship.

As seen in Figure 21, the Wiseman engines have shown increase in torque as the cylinder displacement increased. The trend has a strong correlation ($R^2 > 0.99$), and so the torque seems to be directly proportional to the displacement. As the cylinder displacement of the Wiseman engine increases, the output toque also increases. The Wiseman engine is known to produce lower torque compared to a slider-crank engine of the same size because of its longer stroke length. But in the Wiseman engine as the side loads is negligible, the decrease in torque is compensated by its lowered cylinder friction. Thus, the net torque at output shaft is more.

Once again, the same trend can be noticed in the case of slider-crank engines, from Figure 22. Similar to the Wiseman engines, the output torque of the slider-crank engines also increase as the engine displacement increases. Slider-crank engines running on gasoline, E85, and diesel produce more torque as engine displacement is increased.

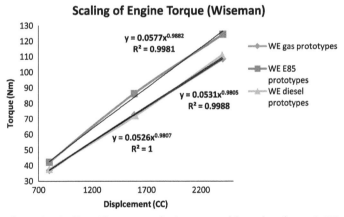

Figure 21. Scaling Wiseman engine's torque with engine size at 2,000 rpm.

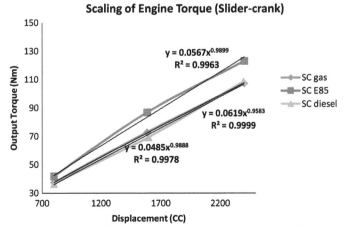

Figure 22. Scaling slider-crank (SC) engine's torque with engine size at 2,000 rpm.

Figure 23 shows a comparison of the output torque of the Wiseman engines and slider-crank engines of the same size, while operating on different fuels. It can be seen that as the engine displacement increased, the Wiseman prototypes produced *slightly more torque* than similar slider-crank engines. The original 30cc Wiseman engine produced less torque than the slider-crank engine because its piston was set higher, providing less combustion volume.

III. Scaling the engine peak power output

Similar to the relationship between torque and displacement, the peak output power of the engine also increased as the engine's combustion volume increased. Though, the change in output power with respect to the displacement was more for the two-stroke engines than the four-stroke engines. It has been theoretically known that the two-stroke engines tend to produce more power per unit displacement (Menon, 2010).

Figure 24 confirms this relationship as it can be seen that the output power of the engine, as a function of the engine displacement, increased as the displacement increased. This is true irrespective of the fuel being used. Figure 25 shows that this is also true for the slider-crank engines of same size. Though, the curves do not follow the same shape and trend as the Wiseman engines. Thus, it can be noted that the Wiseman engines produced more power than the slider-crank engines of the same size.

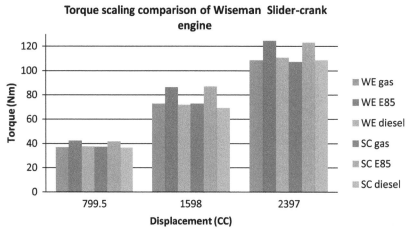

Figure 23. Comparing the scaled torque of Wiseman (WE) and slider-crank (SC) engine.

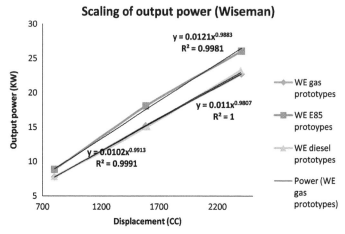

Figure 24. Scaling Wiseman engine's peak power with size at 2,000 rpm.

When comparing the power generated by the Wiseman engine and the slider-crank engine of similar displacements, *it was noticed that the Wiseman variants produced more power than its slider-crank counterparts irrespective of the fuel type.* This can be seen in Figure 26. This was especially true for the 30 HP (2,397cc) Wiseman engine operating on E85.

IV. Scaling the engine fuel consumption at peak power output

The miniature engines that Dr. Menon tested showed decrease in fuel efficiency as the size of the engine decreased. This was due to increase in motoring loses when the engine components were scaled down. Though, for the engines of moderate-to-large sizes, the fuel efficiency increased with increase in displacement. The change in this trend was found after the ranges of 15–20cc (Menon, 2010).

As seen in Figure 27, the Wiseman prototypes showed similar trend, but not with strong correlation considering the lower values of R^2. It can be seen that the fuel efficiency varies with the change in the displacement but overall, there is a slight improvement with increase in the engine displacement. Similar to the results from previous multi-fuel tests, the E85 engines in general consume more fuel than the gas and diesel variants. This could again be due to the lower calorific value of the E85 fuel. Again, the diesel Wiseman engines are most fuel efficient, with the 30 HP variant having the highest efficiency. It was also noticed that the gas variants did not show any significant change in the fuel efficiency with change in displacement.

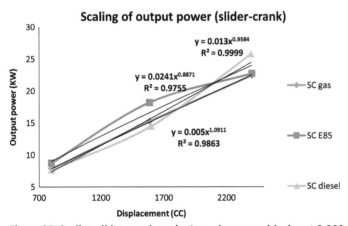

Figure 25. Scaling slider-crank engine's peak power with size at 2,000 rpm.

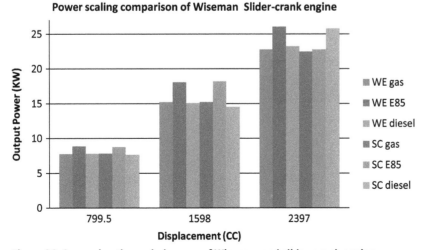

Figure 26. Comparing the scaled power of Wiseman and slider-crank engine.

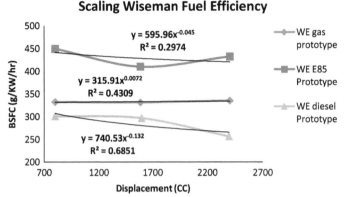

Figure 27. Scaling Wiseman engine's BSFC with respect to engine displacement.

When it comes to the slider-crank versions of the engines, the trend remained the same, but they had slightly lower fuel consumption as the engine displacement increased as shown in Figure 28. This observation was further compared to see how the engines with Wiseman mechanism perform in terms of fuel efficiency when compared to the slider-crank versions of the same size.

By comparing the BSFC results of both the mechanisms in Figure 29, it can be seen that the Wiseman engines tend to be slightly less fuel efficient but the difference is minor. However, both the engines follow a common trend, that is, increase in fuel efficiency with increase in the engine

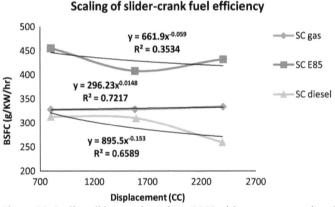

Figure 28. Scaling slider-crank engine's BSFC with respect to engine size.

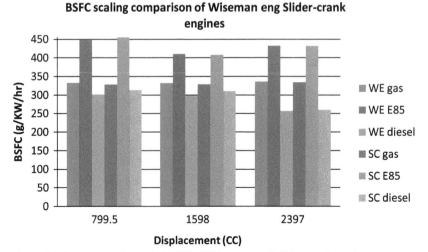

Figure 29. Comparing the scaled BSFC of Wiseman and slider-crank engine.

displacement. The gas and E85 variants of both the engines have almost identical fuel consumption, whereas the Wiseman diesel engines prove to be a slightly more fuel efficient than its slider-crank counterparts.

V. Scaling the engine BMEP

Another important engine parameter to gage the engine's performance is the BMEP. The BMEP decides the work done by the piston, which in turn determines the output torque and power. This reflects on the overall quality of the engine design. It was established that the engine BMEP gradually increases with increase in engine displacement, in majority of the cases (Menon, 2010).

The results in Figure 30 show a similar trend in the peak BMEP of the Wiseman engines. Although the trend is not clearly recognizable due to the small size of the data-set, overall the BMEP seems to increase with increase in cylinder displacement. The established scaling law seems most evident in the gas variants of the engines, whereas the diesel engines tend to show the opposite pattern. Also, the engines operating on E85 fuel produced more BMEP than the gas versions in general, with the 30 HP engine having the highest BMEP output. This pattern was also evident while comparing the performance of the Wiseman 30cc engine operating on gas and E85. Though, the 20 HP ethanol engine had a lower cylinder BMEP than the 20 HP diesel engines.

While scaling the slider-crank versions of the same engines, a different trend was noticed as shown in Figure 31. The engines operating on gas and E85 had a gradual increase in the cylinder BMEP, while the diesel variants followed a trend more similar to the E85 version of the Wiseman engine. Though, the ethanol engines still produce higher BMEP than the gas counterparts. And, the 30 HP version of the diesel engine produces highest BMEP of them all.

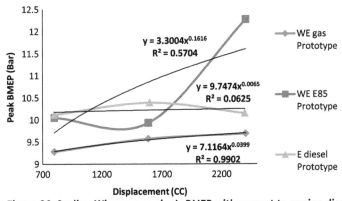

Figure 30. Scaling Wiseman engine's BMEP with respect to engine displacement.

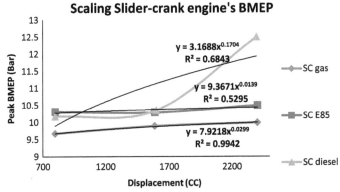

Figure 31. Scaling slider-crank engine's BMEP with respect to engine size.

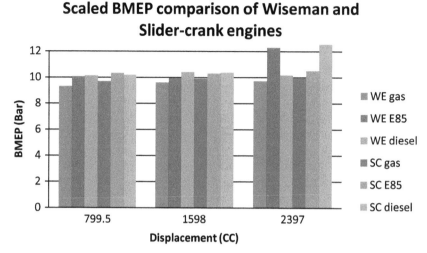

Figure 32. Comparing the scaled BMEP of Wiseman and slider-crank engine.

After comparing the BMEP results of the Wiseman and slider-crank engines in Figure 32, it was observed that slider-crank engines have a higher BMEP than the Wiseman engines in general. There are two exceptions to this trend, and both are for the 30 HP variants of these engines. The 30 HP slider-crank diesel and 30 HP Wiseman ethanol variants produce almost identical BMEP. The BMEP of these two engines is also the highest among all the other engines.

4. Conclusion
In this research, performance of the WHE was examined with respect to power output, torque, and BSFC and compared to similar slider-crank engines. The WHE was modeled in LES using a user-defined subroutine that simulated the hypocycloid mechanism. WHE alpha prototype was modeled and simulated in LES. This prototype was also tested using a dynamometer. The LES simulation results and dynamometer results were comparable and showed similar trend.

Furthermore, a four-stroke diesel version of the Wiseman engine was modeled in the LES software. The simulation results for this engine were promising in almost all aspects. It was seen that this engine produced higher power and torque than the two-stroke Wiseman and slider-crank engines running on gas.

The multi-fuel capability of the Wiseman engine with contra piston was also studied which showed that compared with gasoline, it produced higher power and torque while operating on E85 flex-fuel. To understand the effect of scaling on WHE, four-stroke Wiseman engines were designed to produce 10, 20, and 30 HP, and performance analysis was conducted to predict how the output power, torque, and fuel efficiency would change according to the displacement while operating on different fuels. Overall, it was noticed that the Wiseman engines fared better in terms of power and torque than the slider-crank engines as its size increased. A similar trend was also noticed when fuel consumption was studied. As the Wiseman engine's displacement increased, it was found to be more fuel efficient due to reduced motoring losses in it. It was found that the 30 HP Wiseman engine, while operating on diesel, had the best fuel efficiency. This trend in increased performance with increase in displacement was also true for the cylinder BMEP.

The scaling and multi-fuel analysis suggested that four-stroke diesel and ethanol variants of the Wiseman engine were very promising compared to the two-stroke gasoline engines. Currently, work is underway to prototype Wiseman engines studied in this research. Later, these engines will be tested using a dynamometer and experimental results will be compared with the simulations presented in this paper.

Funding

This research was partially supported by the funding from Science Foundation Arizona and Wiseman Technologies Incorporated.

Author details

Priyesh Ray[1]

E-mail: priyesh.ray@asu.edu

Sangram Redkar[1]

E-mail: sredkar@asu.edu

[1] Department of Engineering, Arizona State University, Mesa, AZ 85212, USA.

References

Andriano, M. B. (1998). *Design, construction and testing of hypocycloid machines*. Detroit: SAE International.

Annand, W. J. D. (1963). Heat transfer in the cylinders of reciprocating internal combustion engines. *Archive: Proceedings of the Institution of Mechanical Engineers 1847–1982 (vols 1–196), 177*, 973–996. doi:10.1243/PIME_PROC_1963_177_069_02

Beachley, N. H., & Lenz, M. A. (1988). *A critical evaluation of the geared hypocycloid mechanism for internal combustion engine application*. Detroit, MI: Society of Automotive Engineers. http://dx.doi.org/10.4271/880660

Cahyono, B., & Bakar, R. A. (2010). Effect of ethanol addition in the combustion process during warm-ups and half open throttle on port-injection gasoline engine. *American Journal of Engineering and Applied Sciences, 4*, 66–69. doi:10.3844/ajeassp.2011.66.69

Celik, M. B. (2007). Experimental determination of suitable ethanol–gasoline blend rate at high compression ratio for gasoline engine. *Applied Thermal Engineering, 28*, 396–404. doi:10.1016/j.applthermaleng.2007.10.028

Conner, T. (2011). *Critical evaluation and optimization of a hypocycloid Wiseman engine* (Master's thesis). Arizona State University, Mesa, AZ.

Costa, R. C., & Sodre, J. R. (2010a). Compression ratio effects on an ethanol/gasoline fuelled engine performance. *Applied Thermal Engineering, 21*, 278–283.

Costa, R. C., & Sodre, J. R. (2010b). Hydrous ethanol vs. gasoline–ethanol blend: Engine performance and emissions. *Fuel, 89*, 287–293. http://dx.doi.org/10.1016/j.fuel.2009.06.017

Datta, A., Chowdhuri, A. K., & Mandal, B. K. (2012). Experimental study on the performances of spark ignition engine with alcohol–gasoline blends as fuel. *International Journal of Energy Engineering, 2*, 22–27.

Ganesan, V. (2012). *Internal combustion engines* (4th ed.). New Delhi: Tata McGraw.

Hatz Diesel. (2012). *Operator's manual: Diesel engines*. Germany. Retrieved July 21, 2014, from http://www.hatz-diesel.com/uploads/tx_hatzproducts/BA_1B_EN_43380210.pdf

Ishida, K., & Matsuda, T. (1975). Fundamental researches on a perfectly balanced rotation–reciprocation mechanism. *Bulletin of the JSME, 18*, 185–192.

Karhula, J. (2008). *Cardan gear mechanism versus slider crank mechanism in pumps and engines*. Lappeenranta: Lappeenranta University of Technology.

Menon, S. (2010). *The scaling of performance and losses in miniature internal combustion engines* (Doctoral dissertation). University of Maryland, College Park, MD.

Ruch, D. M., Fronczak, F. J., & Beachley, N. H. (1991). 911810 —Design of a modified hypocycloid engine. *SAE International*, 73–89.

Special plane curves. (2005–2014). Retrieved July 21, 2014, from http://www.math10.com/en/geometry/analytic-geometry/geometry5/special-plane-curves.html

Topgül, T., Yücesu, H. S., Çinar, C., & Koca, A. (2006). The effects of ethanol–unleaded gasoline blends and ignition timing on engine performance and exhaust emissions. *Renewable Energy, 31*, 2534–2542. http://dx.doi.org/10.1016/j.renene.2006.01.004

US Department of Energy, Energy Efficiency & Renewable Energy. (2013). *Handbook for handling, storing, and dispensing e85 and other ethanol–gasoline blends*. Retrieved July 21, 2014, from http://www.afdc.energy.gov/uploads/publication/ethanol_handbook.pdf

Wiseman, R. (2001). *US Patent No. 6,510,831*. Waveland, MS: US Patent and Trademark Office.

Wiseman Engine Group. (2010). *The engine of the future–today!* Pass Christian, MS: Business Plan.

Wiseman Technologies Inc. (n.d.). Retrieved July 21, 2014, from http://www.wisemanengine.com/AUVSIWisemanPaPer.doc

7

Effects of disordered microstructure and heat release on propagation of combustion front

Tarun Bharath Naine[1] and Manoj Kumar Gundawar[1]*

*Corresponding author: Manoj Kumar Gundawar, Advanced Centre of Research in High Energy Materials (ACRHEM), University of Hyderabad, Hyderabad 500046, Telangana, India
E-mails: manojsp@uohyd.ernet.in, manoj@uohyd.ac.in
Reviewing editor: Duc Pham, University of Birmingham, UK

Abstract: Numerical experiments for diagnosis of combustion of actual heterogeneous systems are performed on a one-dimensional chain. The internal microstructure of actual heterogeneous systems is apriori unknown, various distributions like uniform, beta, and normal have been considered for distributing neighboring reaction cells. Two cases, for the nature of distribution of heat release of reaction cells are taken into account, one with identical heat release and the other with disordered heat release. Role of different random distributions in describing heterogeneous combustion process is established in present paper. Particularly, the normal distribution of arranging neighboring reaction cells has been found to be powerful methodology in explaining the combustion process of an actual heterogeneous system at higher ignition temperatures for both cases of distributing heat release. Validation of the developed model with the experimental data of combustion of the CMDB propellants, gasless Ti + xSi system, and different thermite mixtures is performed. Our results show that the experimental burning rates at higher ignition temperatures ($\varepsilon > 0.32$) of the heterogeneous system are better reproduced theoretically with the present model. We have also shown that different combustion limits for different thermite systems are the consequences of disordered heat release. Experimental data for thermite systems that have lower inflammability limits are

ABOUT THE AUTHOR

Manoj Kumar Gundawar is working as an assistant professor at university of Hyderabad, Hyderabad, India. His primary research interests are experimental work on laser-induced breakdown spectroscopy. His group is also interested in the modeling aspects of the heterogeneous combustion, particularly, randomizing the differences in positions of the combustion cells. The most important discovery from his group shows that the random structure of the microheterogeneous system plays a crucial role in the dynamical and statistical behavior of the system, a direct consequence of the nonlinear interaction of the structure of the system with the thermal wave. Present manuscript further extends the idea of random distributions to other distributions which is apriori unknown parameter. The experimental burn rates can be associated with one of the distributions by matching them with the numerical burn rates. This kind of strategy enables the predication of the burn rates for future studies and different amounts of mixtures.

PUBLIC INTEREST STATEMENT

Modeling and simulation of heterogeneous combustion can overcome the difficulties of experimental investigations, has inherent safety and wide operational features. The actual heterogeneous mixtures while packing involves improper mixing, percentage of diluters, and binding agents. These situations have their own impact on internal microstructure, modes of combustion front, and combustion limit. The present manuscript explains the effect of these situations on combustion process by considering various distributions like uniform, beta, and normal for distributing neighboring reaction cells. Two cases, for the nature of distribution of heat release of reaction cells are taken into account, one with identical and the other with disordered heat release. Our results show that the normal distribution of arranging neighboring reaction cells can be the powerful methodology in explaining the combustion process of an actual heterogeneous system. We have also shown that different combustion limits for different thermite systems are the consequences of disordered heat release.

analyzed in the view of disordered heat releases of cells. The model developed in the view of disordered heat releases reproduces the experimental burn rates and experimental combustion limit.

Subjects: Computational Physics; Physical Sciences; Physics; Statistical Physics; Theoretical Physics

Keywords: heterogeneous combustion; heat release; ignition temperature; burn rate; normal distribution

1. Introduction

Combustion of powdered mixtures (Beckstead & McCarty, 1982; Bharath, Rashkovskiy, Tewari, & Gundawar, 2013; Denisyuk, shabalin, & Shepelev, 1998; Dvoryankin, Strunina, & Merzhanov, 1985; Kubota, 1978, 2002; Kubota & Okuhara, 1989; Kulkarni & Sharma, 1998; Rashkovskii, 2005; Rashkovskiy, Kumar, & Tewari, 2010; Rogachev & Baras, 2009) and burning of solid propellants, liquid droplet combustion, spray combustion, combustion of coal, and engines are the examples of heterogeneous combustion. CMDB (Kulkarni & Sharma, 1998) propellants (DNC + CL, DNC + CL + AP), also known as composite modified double base, are heterogeneous mixtures of solid components and small particulate oxidizers held together with a rubbery material referred to as binders (AP). Such systems with addition of binding agent (AP) become high-performing propellants and are realized with minimum release of gasses. The combustion efficiency increases with an increase in the binding agents. Powdered systems with gasless combustion are as well realized as the simplest form of heterogeneous system. Heterogeneous combustion of the mixtures such as Ti + xSi (Rogachev & Baras, 2009), Ni + xAl commonly involves in the formation of gasless combustion products. Chemical reaction, ignition, and burning away of such mixtures occur in solid state, and result in solid combustion products. Furthermore, combustion of thermite systems such as Fe_2O_3 + Al, Fe_2O_3 + Ti, Cr_2O_3 + Zr (Dvoryankin et al., 1985) are apparently linked with the formation of gaseous combustion products. Fortunately, for the above-mentioned thermite systems, a gasless combustion can still be obtained by diluting the respective thermite mixtures with their respective end product dilutent (Al_2O_3, TiO_2, ZrO_2). Self-sustained combustion for such systems is realized within the limits of quantity of inert dilutent. Self-sustained high-temperature synthesis which is used for the synthesis of materials is an important class of heterogeneous combustion.

The behavior of combustion front in such heterogeneous mixtures, with and without addition of binding agents, and at different stoichiometric mixtures of reactants, also at limits of combustion for different reactants, is of fundamental interest in explaining the combustion process through most accurate model. Such models can account for the effects of practical synthesis of materials, percentage of binding agent, stoichiometries, and inert dilutent on the combustion process. Despite extensive investigations (Bharath et al., 2013; Rashkovskii, 2005; Rashkovskiy et al., 2010), different combustion limits for the thermite mixtures (Fe_2O_3 + Al, Fe_2O_3 + Ti, ...) and modes of combustion front in heterogeneous mixtures are unexplained because of the internal microstrucuture of actual heterogeneous mixtures is apriori unknown and absence of accurate models in agreement with experimental data. The role of internal microstructure and consumption of reactant (distribution of heat release) in combustion process became apparent after numerical (Bharath et al., 2013; Rashkovskiy et al., 2010) and experimental studies using high-speed micro video camera. Such studies have shown that the self-propagating combustion front becomes complex waves at macro scales. The HSMV camera revealed (Willcox, Brewster, Tang, & Stewart, 2007; Favier, 2004; Gardner, Romme, & Turner, 1999; Hwang, Mukasyan, & Varma, 1998; Kerstein, 1987; Mukasyan & Rogachev, 2008; Rashkovskii, 1999; Rogachev, 2003; Rogachev & Baras, 2007; Viegas, 1998) the mechanism modes of combustion which is either quasi-homogeneous or relay race. Heterogeneous combustion is complicated by microstructure and the consumption of reaction cells of the system even at stationary mode of combustion. Such complications results in micro-oscillations of combustion wave and leads to limit of combustion front propagation. Combustion limit is a situation where the combustion front doesn't propagate further. Combustion limit is different for different reactants and the analysis for combustion limits is less concentrated in earlier literatures (Bharath et al., 2013; Rashkovskii, 1999, 2005; Rashkovskiy et al., 2010). Such an analysis for extreme reaction condition during gasless combustion process requires new methods for tailoring the internal microstructure of the system (Candel, 2002; Goroshin, Lee, & Shoshin, 1998; Gross, 2010; Humphrey, 1971; Varma & Mukasyan, 2001). Actual heterogeneous mixtures are of three dimensional ones, the internal microstructure is automatically formed while mixing and packing and the particles in them have definite size depending on concentration of diluents. As demonstrated in Bharath et al. (2013), Rashkovskii (2005), and Rashkovskiy et al. (2010), the one-dimensional study of heterogeneous combustion process can still explain the role of different physical parameters (internal micro structure, heat release, burn rate, disordereness) if the particles are considered of point size. In Bharath et al. (2013), we have not only established and validated the relevance of one-dimensional model but also have shown that the Arrhenius macrokinetics observed during actual combustion process can be related with ignition temperature (T_{ign}). The one-dimensional model has purely thermal nature. However, the work shown in Bharath et al. (2013) did not concentrate on different aspects such as the role of different random distributions and randomness of heat release of reactive cells.

In this study, we concentrate on diagnosis of combustion with significance to two factors that are different and extensions to our previous work (Bharath et al., 2013) (a) different random distributions employed for positioning neighboring reaction cells (b) Randomizing the heat release by considering a distribution of heat release among all the unburnt cells. The methodology, resembles to Monte Carlo simulation, enable us to perceive the affect of minute changes in microstructure on dynamic combustion parameters such as stability of combustion front, ignition delay times, and average burn rate. Such minute aspects are controlled by our model and are illustrated with significant change in parameters such as scale and shaping parameter of distribution. The affect of randomizing heat releases on combustion process is given very little importance and atmost times neglected in earlier literatures. Such situations commonly arise from improper mixing of heterogeneous mixtures, changing the diluter and percentage of diluter. Two different cases of reactant consumption are considered, in present work, by identical distribution of heat release or by randomizing the heat release. The above problem is addressed by describing the system in terms of the other random distributions (normal, beta and uniform). Particularly, the normal distribution allows for positioning the heat sources with minimum clusterization and symmetric around the mean. The novel approach shall optimistically list the priority of distributions in modeling a discrete combustion process.

Section 3 reports the numerical results obtained for developed system. The burn rate of the system including the dynamical behavior of reaction waves in such systems is reported. The Possible limits of a combustion process are studied for normal distribution. Sections 3.1–3.5 discusses the numerical results obtained for identical heat release. The heat release is carried through thermal bridge and its physical significance is explained in modeling section. The developed model is validated and compared with experimental results performed on combustion of CMDB propellants, Ti–Si system, and different thermite systems. Here we establish the most accurate distribution of neighboring reaction cells and its physical significance. Most works on combustion modeling have reported that the effect of random heat release in combustion process (Bharath et al., 2013; Rashkovskiy et al., 2010) is negligible and considered the heat release to be identical for all the cells. Section 3.6 studies the systematic effect of heat consumption of reaction cells (Gardner et al., 1999; Hwang et al., 1998; Mukasyan & Rogachev, 2008; Rogachev, 2003) on combustion process.

2. Modeling

During combustion process, the exothermic reaction is restricted to a thin zone that propagates throughout the combustible system. The vicinity of the reaction zone is referred as combustion (reaction) front. The heat transfer takes place across the entire combustor volume. Traditional explanation of gasless combustion process is based on diffusion process of reaction cell that undergoes exothermic reaction. During practical situations, sample reacting generally has cylindrical geometry. Temperature gradient along transversal direction is very little under many circumstances. Neglecting heat loss upon such situations the one-dimensional model, within the macroscopic framework, can be used for describing the temperature profiles. One-dimensional discrete combustion model, similar to our previous works (Bharath et al., 2013; Rashkovskiy et al., 2010), comprising chain of reaction cells (point particles) distributed either in periodic or random trend is considered here. Nevertheless, in the present methodology, we not only introduce different random distributions (uniform, beta, normal) for spacing neighboring reaction cells but also randomize the consumption of reaction cells. The instabilities of heterogeneous combustion process can be predicted in the framework of macroscopic studies. The statistical study of instabilities, significant to different minute aspects of random process, allows us to identify the situation at macroscopic levels. Combustion process, at higher ignition temperatures, is of complicated in nature. Detailed work in this direction leads us with the possibility of finding the accurate model of heterogeneous combustion process at higher ignition temperatures. Propagation of combustion front is modeled by distributing immobile reaction cells on one-dimensional chain and taking into consideration of spontaneous combustion of reaction cell with heat release (Q_i) as soon as temperature of the active cell reaches a pre-determined ignition temperature (T_{ign}). Experimental data (Dvoryankin et al., 1985; Kulkarni & Sharma, 1998; Rogachev & Baras, 2009) illustrate that the chemical reaction of the mixtures begins after the melting of reactants. Ignition and burning time for active cells in such systems are always less when compared to characteristic time for heating of reaction cells. These observations during the experiments permit us to regard the combustion process of active cells in the system as spontaneous. The moment at which the threshold temperature (T_{ign}) is reached (melting of reactants begin) is considered as ignition time of the reaction cells. The chains of reaction cells described in the model are interconnected using a thermal bridge. Thermal bridge is described by thermal conductivity k, linear mass density ρ, and specific heat c. Heat transfer between reaction cells is carried through the thermal bridge and it is assumed that there are no heat losses i.e. total heat energy released by combustion of preceding particle is utilized completely for combustion of succeeding unburnt particles. The model under consideration simplifies the reaction kinetics of actual heterogeneous combustion process. The parameters of developed model such as ignition temperature and heat release (consumption of reactant) signify the characteristic features of the reactants. If we regard this model to actual combustion of heterogeneous mixtures (Dvoryankin et al., 1985; Kulkarni & Sharma, 1998; Rogachev & Baras, 2009), the sense of consumption of reactant (heat release) and ignition temperature becomes apparent with respect to combustible limits and quantity of inert diluent. The gasless combustion of heterogeneous mixtures, chemically inert by itself, undergoes self-sustained combustion after releasing a large amount of heat. Such process of combustion is achieved only after initial melting of reactants. In the present model, sufficient amount of initial thermal energy is given to the system by initiating a fraction of cells at time ($t = 0$). The problem in consideration is explained by a one-dimensional equation of thermal conductivity of point cells as (see Rashkovskiy et al., 2010):

$$T(t, x) = T_{in} + \sum_{i(t)} \Delta T_i(t - t_i, x - x_i),$$

where T_{in} is the initial temperature of the system; t_i is the instant of ignition of the ith cell, located at x_i. The above equation has a solution given by Equation (1). Governing equation for a one-dimensional combustion process involving dimensionless parameters (Bharath et al., 2013; Rashkovskiy et al., 2010) is as follows:

$$2\sqrt{\pi\varepsilon} = \sum_{i=-\infty}^{k-1} \frac{q_i}{\sqrt{s_k - s_i}} \exp\left(-\frac{(\xi_k - \xi_i)^2}{4(s_k - s_i)}\right) \tag{1}$$

In the above, the non-dimensional parameters are: ignition time for burning cell $s = t(k/l_o^2)$, distance of unburnt particle $\xi = x/l_o$, ignition temperature $\varepsilon = \frac{T_{ign} - T_{in}}{T_{ad} - T_{in}}$ and heat release in combustion $q_i = Q_i/Q_o$ (here in this paper, we consider two cases of q_i one with identical value at all heat sources and other with randomized), where $T_{ad} = T_{in} + \frac{Q_o}{\rho c l_o}$ is mean adiabatic temperature of the thermal bridge, $Q_o = \lim_{N \to \infty} \frac{1}{N} \sum_{i=1}^{N} Q_i$ is a mean energy release in combustion of one heat source and $l_o = \lim_{N \to \infty} ((x_N - x_o)/N)$ is a mean distance between neighbor point heat source of the system.

Equation (1) is utilized for the numerical calculation of the ignition times of all particles in the developed model (system). The model developed here is described by a single parameter ε (non dimensional ignition temperature of active cells). Spacing of unburnt reaction cells in periodic system is done by regular spacing at a distance of unit. Random system that involves distribution of positioning unburnt cells is achieved by probability density function of various random distributions, namely uniform, normal, and beta with a mean of unity. The probability density function for different distributions is listed in Table 1. A gamma distribution with shaping parameter $a = 0.7$ (Bharath et al., 2013) is also considered for comparison. Total reaction cells in the model are considered as concatention of 1,000 burnt and 2,500 unburnt cells. Figure 1 shows the probability density functions for spacing unburnt cells significant to shape and scale. The average spacing of particles is always maintained to be unity to facilitate a direct comparison of the results with a periodic system (Rashkovskii, 2005).

The combustion process for such systems (disordered spacing of reaction cells) takes place non-uniformly in oscillating mode (see Bharath et al., 2013; Rashkovskiy et al., 2010); hence average burn rate is considered. The average burn rate of combustion front in the system between cells i and $k > i$ is equal to:

$$\omega_{\text{\tiny IK}} = (\xi_{\text{\tiny K}} - \xi_{\text{\tiny I}})/(S_k - S_i) \tag{2}$$

The numerical simulation of the modeled system requires high-computing speed. C language, with the facility of MPI, for simulation of combustion process is employed for executing the calculation.

S. no.	Distribution	Probability density function	Parameters	Mean
Table 1 Probability density function for different random distributions				
1.	Normal	$p(L) = \frac{1}{\sigma\sqrt{2\pi}}e^{\frac{-(L-\mu)^2}{2\sigma^2}}$	σ, μ	μ
2.	Uniform	$p(L) = \frac{(L-a)}{(b-a)}$	a, b	$\frac{a+b}{2}$
3.	Beta	$p(L) = \frac{1}{B(a,b)}L^{(a-1)}(1-L)^{(b-1)}$	a, b	$\frac{a}{a+b}$

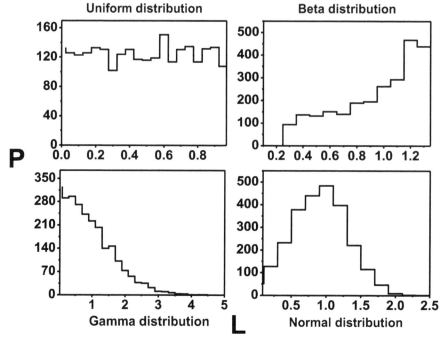

Figure 1. Probability density function for different distributions.

3. Numerical results

3.1. Identical heat release

Ignition time profiles are obtained using Equation (1) for different distributions of positioning neighboring reaction cells at different ignition temperatures ε. Note; here we consider identical distribution of heat release at all reaction cells. Figure 2 shows the ignition time profiles at different ignition temperatures of a combustible system described by different probability density functions employed for positioning neighboring reaction cells. $\sigma = 0.05$ for normal distribution and $a = 0.7$ for gamma distribution are considered for comparison. Ignition time profiles depend both on the ignition temperatures and minute aspects of distributions of neighboring reaction cells. The randomness (Bharath et al., 2013; Rashkovskii, 2005; Rashkovskiy et al., 2010) in microstructure of a system also has critical influence on the ignition time profiles (Rogachev & Baras, 2009). The ignition time profiles are linear for both normal distribution and beta distribution of neighboring reaction cells and are identical to the periodic system. This can be inferred from Figure 1, where it can be seen for a beta distribution about ~70 percent of the cells spacing's are in the range from 0.8 to 1.2 and ~100 percent for the normal distribution. The particles in spite of their random spacing bear a close resemblance to a periodic system with unit spacing and hence a linear relation. Such an analysis determines the effect of internal microstructure (packing of particles) on combustion process. However, the combustion front, for gamma distribution of neighboring reaction cells with shaping parameter $a = 0.7$, is noticeable with more stops as the ignition temperature increases. Also a similar affect can be seen for a normal distributed system, which will be shown in next Section. Instability in combustion or reaction front stops (Bharath et al., 2013; Gardner et al., 1999; Hwang et al., 1998; Mukasyan & Rogachev, 2008; Rashkovskii, 2005; Rashkovskiy et al., 2010; Rogachev, 2003; Rogachev & Baras, 2009) is more prominent at higher ignition temperatures and higher degree of randomness in microstructure of the system. At higher ignition temperatures, though the quantity of heat energy obtained from the burnt cells is same as compared to a corresponding periodic system, the combustion front sees stops in the propagation as a result of the randomness of neighboring reaction cells. With an increase in the degree of randomness of positioning neighboring reaction cells, the obtained heat energy has to overcome the disordered microstructure of system, particularly those points on the chain where the neighbor-cell spacing is very large (Gardner et al., 1999; Hwang et al., 1998; Mukasyan & Rogachev, 2008; Rogachev, 2003). This can be quantified by observing the maximum ignition delay times of a system. The stops in combustion front are result of increased ignition delay times. Delays in ignition times are due to increase in induction periods between active cells and ignition temperature of a combustible system.

Figure 3(a) shows plot between ignition temperature and maximum ignition delay times of a system. These maximum ignition delay times are obtained from the correlation analysis between adjacent reaction cells and their ignition delay times.

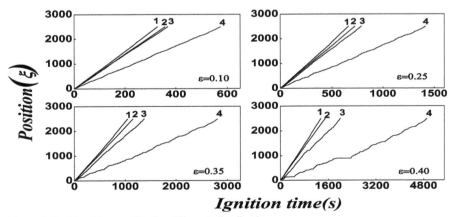

Figure 2. Ignition time profile for different distributions.

Notes: (1) Normal (2) Beta (3) Uniform (4) Gamma ($a = 0.7$).

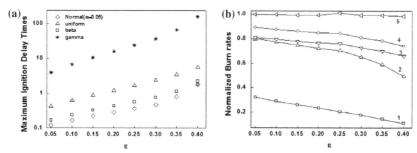

Figure 3. (a) Plot for ignition temperature and maximum ignition delay times of different distributions. (b) Plot for ratios of random to periodic burn rates.

Notes: (1) Gamma ($a = 0.7$) (2) Normal ($\sigma = 0.4$) (3) Uniform (4) Beta (5) Normal ($\sigma = 0.05$).

From the correlation analysis, it is observed that the variance of the ignition delay times increases with an increase in the degree of randomness in microstructure and ignition temperature. From Figure 3(a), it is observed that maximum delay in ignition times of a system is proportional to degree of randomness in microstructure and ignition temperature of the system. The maximum delay in ignition times at particular ignition temperature shows the trend periodic < normal ($\sigma = 0.05$) < beta < uniform < gamma ($a = 0.7$). The maximum delay in ignition times also plays the critical role in effecting the stability of combustion front and thus burn rates. Burn rates of the systems are obtained using Equation (2) and are normalized to burn rates of a periodic system for different ignition temperatures as shown in Figure 3(b). Figure 3(b) shows that burn rates of a Periodic system can be reproduced approximately by random systems described with normal ($\sigma = 0.05$) distribution. Again this is a consequence of closeness of these systems to a periodic system and is explained in the theoretical analysis section. By changing the parameters of a distribution, the inter cell spacing can be altered, which is reflected in the probability density function. For beta and uniform distributions, we observe no change in shape of distribution and consequently there is no effect on burn rates. However, for normal distribution, the burn rates can be varied by changing the parameter (σ). The normalized burn rate obtained with disordered system described by normal distribution with minimum variance is observed to be ~98% at each ignition temperature but for system's microstructure described by uniform, beta distributions the percentage of normalized burn rates decreases with an increase in the ignition temperature. As the nature of distribution of reaction cells in actual systems is apriori unknown and hence it is of academic interest to find out the accurate model. Compared to gamma distribution (Bharath et al., 2013) of neighboring reaction cells, the normal distribution of reaction cells also has an advantage of parameter known as standard deviation. Hence the numerical calculations for model developed in the view of normal distribution of neighboring cells have to be carried out to establish accurate model and also explore the limits of different models. Normal distribution is employed for modeling in this paper as it offers several advantages—its fixed shape and positioning adjacent cells with controlled deviation from mean, around 65 percentage of adjacent reaction cells are positioned within one standard deviation, describe the microstructure of high energy material (Kulkarni & Sharma, 1998) mixtures. The flexibility of using normal distribution is due to fact that curve may be centered over any number of real lines and it may be flat or peaked to correspond to amount of dispersion in the values of random variable. The variance of normal distribution corresponds to concentration of point source particles of system when compared to that of periodic system. Different cases of normal distributed systems characterized with different standard deviations (~0.05, 0.19, 0.28, 0.40, ...) are considered for modeling and simulation. The combustion stops are more prominent for higher values of variance and higher ignition temperatures. The induction periods for higher ignition temperatures and higher variance of adjacent reaction cells increase. Symbols in Figure 4 show the normalized reaction burn rates calculated for different normal distributions. This detail provides us the possibility of tailoring burn rates of a system by suitable normal distribution. The normalized burn rates of a normal distribution are in the wide range (0.3–1). It can be seen that ratio = 1 resembles to a periodic system and 0.3 is closer to a high degree of disordered system. As variance of positioning adjacent cells in developed system increases, the concentration of sample density (Rogachev & Baras, 2009) decreases and its reaction front propagation corresponds to relay race homogeneous mechanism where as the reaction front propagation for the system with low variance corresponds to quasi homogeneous mechanism (Gardner et al., 1999; Hwang et al., 1998; Mukasyan & Rogachev, 2008; Rogachev, 2003; Rogachev & Baras, 2009). Our numerical simulations suggest that apart from gamma distribution of reaction cells, the normal distribution of reaction cells also play a crucial role in describing the wide range of the combustion of actual heterogeneous systems. This shall be made evident by comparing the developed model (utilizing normal distribution of neighboring reaction cells) with available experimental data. In due process of comparison, we also try to establish scopes of the developed model.

3.2. Theoretical analysis for burn rates

The present section illustrates the theoretical analysis for dependence of burn rates of a disordered system, normalized to burn rates of periodic system, on variance of the distribution of neighboring reaction cells. Thus obtained theoretical expression is also compared with numerical results. Burn rates of combustible system (developed model) under consideration are determined by sequential calculation of ignition times of reaction cells. In the view of academic interest, it is essential to perform the correlation analysis between delay in ignition times and neighboring reaction cells. The correlation analysis is performed similar to the description as shown in theoretical section (Rashkovskiy et al., 2010). The non-dimensional steady-state burn rate of a periodic system can be written as:

$$\tau = \omega_p^{-1} L \tag{3}$$

where τ is the delay in ignition times for neighboring reaction cells; L is the non-dimensional distance between neighboring reaction cells; $\omega_p(\varepsilon)$ is the non-dimensional burn rate of periodic system where neighboring reaction cells are positioned with unit distances. The correlations (m), between τ and L, are obtained for a disordered system, described by a normal-distribution of neighboring reaction cells (L), as a functional dependence on variance (σ^2) of distribution of neighboring reaction cells (L) and ignition temperature (ε). As $\sigma^2 \to 0$, the disordered system described by normal distribution of neighboring reaction cells (L) is similar to periodic system. Correlation analysis is performed on numerical calculations of Equation (1) for a disordered system over a broad range of parameters $\sigma^2 \in [0.4 \ldots 0.05]$ and $\varepsilon \le 0.48$. The correlation factor (m) thus obtained in terms of power functional dependence of $m(\varepsilon, \sigma^2)$ is as follows:

$$m = 2 + 0.2 \exp\left(3.18\varepsilon^{0.66} + (0.1 + 0.029 \ln \varepsilon)\sigma^2 - (0.0012 + 0.003\varepsilon)\sigma^4\right) \tag{4}$$

As $\sigma^2 \to 0$ for a disordered system, the correlation limit is $m \to 2$ at $\varepsilon \to 0$. The limiting range considered for normal distribution of neighboring reaction cells (L) is $\sigma^2 \in [0 \ldots 0.4 \ldots 0.05]$ hence Equation (4) can be further modified to straightforward expression as:

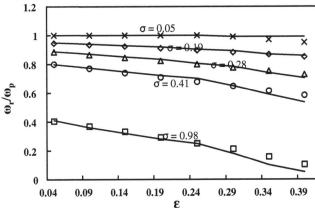

Figure 4. Comparison plot between normalized burn rate and nondimensional ignition temperature ε for different values of σ. Markers are obtained using Eqs. (1) and (2); solid lines are obtained by the theoretical Eqs. (5), (6) with $\phi = 1$.

$$m = 2.2 - 0.01(1/\sigma^2) + 2.23(1/\sigma^2)^{0.27}\varepsilon \tag{5}$$

The expectation value for L^m with L described by normal distribution in accordance with periodic system Equation (3) is given as:

$$\langle L^m \rangle = \int_0^\infty L^m P(L)dL$$

The above equation converges for m being integer and the solution obtained is given below as:

$$\omega_r/\omega_p = \frac{1}{\varphi}(1/\sigma^2)^m \frac{\Gamma(1/\sigma^2)}{\Gamma((1/\sigma^2) + m)} \tag{6}$$

In the limit $\sigma^2 \to 0$, the system becomes periodic; this shows that $\omega_r \to \omega_p(\varepsilon)$, in accordance to Equation (3) $m \to 2$ in this limit. Subsequently, we obtain $\lim_{\sigma^2 \to 0} \frac{(1/\sigma^2)^m \Gamma(1/\sigma^2)}{\Gamma((1/\sigma^2)+m)} = 1$; thus one can conclude that $\phi \to 1$ for this limit. Figure 4 shows the comparison of burn rates, obtained by direct numerical solution of Equations (1) and (2), represented by markers, and the theoretical dependency, calculated using Equations (5) and (6), represented by solid lines, with $\phi = 1$.

It is evident that the theoretical dependence Equations (5) and (6) with $\phi = 1$ properly explains the results calculated by numerical simulations using Equations (1) and (2) for the whole range of parameters σ. Generalized expression for the theoretical dependence of normalized burn rates on variance is obtained and now can be used for direct calculation of burn rate if σ is known.

3.3. CMDB Propellants

In this section, we compare experimental data for CMDB class of propellants (Kulkarni & Sharma, 1998) with the model developed. Systems modeled with normal or gamma distribution of neighboring reaction cells are compared with experimental data (Kulkarni & Sharma, 1998) to establish accurate model and also account the affects for dynamical combustion properties. In the experiments of Kulkarni and Sharma (1998), the samples were comprised of selected components such as DNC (Dinitrocarbanilide) and CL mixed in exact ratio with a binding agent AP (Amonium perchlorate) or without binding agent. The burning temperatures and burning rate of such propellants can change over broad range either by adding binding agent (AP) or by changing the compositions of components; the mechanism of heat release during combustion process of such propellant is not altered. In such a system, the group of active cells play the role of reaction cells, while the components are capable of reacting chemically. T_{ign} is the melting point at which the cells commence reacting. The values of burning temperature and burning rate of CMDB propellants with binding agent or without binding agent are collected from the work (Kulkarni & Sharma, 1998). Ignoring the heat losses, calculated burning temperature of propellant mixtures is recognized as adiabatic temperature (T_{ad}) with reference to developed system. Figure 5(a) includes the data from the work (Kulkarni & Sharma, 1998), with burn rate and burning temperature on the axis, for two propellants. The burn rate for CMDB propellants varies with percentage of components and also by adding binding agents. The first propellant comprises 60% of DNC and 40% of CL without the binding agent and other propellant comprises 50% of DNC and 40% of CL with AP as binding agent. It is observed that burn rate decreases with addition of binding agent, particularly at 690 k.

A similar treatment, using Equations (27–29) as shown in Bharath et al. (2013), is performed in the coordinates of $\varepsilon-\omega$ for experimental data (Kulkarni & Sharma, 1998). In doing so it is considered that critical temperature, noticed in experiments, corresponds to critical values of the theoretical parameters such as ε_{cr} and ω_{cr}. In view of the developed model, the critical ignition temperature is established at 0.48; we consider these parameters as critical parameters for the disordered system. Assuming that minimum burn rate and burning temperatures, determined in the experiments,

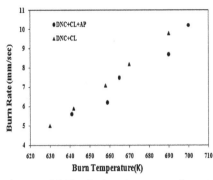 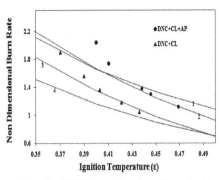

Figure 5. (a) Plot for burning rate vs. burn temperature for CMDB propellants with and without binding agent, based on data of work (Kulkarni & Sharma, 1998). (b) Comparison plot for theoretical and experimental dependencies $\omega(\varepsilon)$. Symbols are converted experimental data (Kulkarni & Sharma, 1998).

Notes: In (b) solid lines 1–4 are the theoretical dependence, lines 2&3 represents normal distributed system: (2) $\sigma = 0.2886$: (3) $\sigma = 0.408$; lines 1&4 represents gamma distributed system:(1) $a = 12$; (4) $a = 6$.

correspond to the critical regime of combustion for real system. The values $\omega_{cr} = 5$ and $\omega_{cr} = 5.6$, respectively, were used for the mixtures DNC + CL and DNC + CL + AP in these calculations.

We assume that the change of burning rate with change in burning temperature for combustion of propellants is associated with change of parameters either σ in case of normal distribution or a for gamma distribution which describes system internal microstructure considering uniform identical reactant consumption. The data (Kulkarni & Sharma, 1998) are now processed in variables $\varepsilon - \omega$. The numerical results calculated for appropriate parameters of σ, a at different ignition temperatures ε, is superimposed in Figure 5(b). Solid lines 1–4 are the theoretical dependences $\omega(\varepsilon)$, calculated using expressions Equations (6), (5) taking into account the matched dependence $\sigma(\varepsilon)$. Line 1 and 4 is obtained by utilizing the gamma distribution of neighboring reaction cells with parameter $a = 12$ and $a = 6$, respectively. Line 2 and 3 is obtained by utilizing the normal distribution of neighboring reaction cells with parameter $\sigma = 0.2886$ and 0.408, respectively.

Combustion of CMDB propellants releases high amount of energy, which propagates throughout the system, even at low burning temperatures. Thermal reactions below 670 k are like streak of lightening and are complex to analyze; however for the burning temperatures above 670 k, the developed model (Line 2 and 3) shows close agreement when compared to that of with Line 1 and 4. The minute aspect of controlling internal micro structure (Candel, 2002; Dvoryankin et al., 1985; Rogachev & Baras, 2009), by normal distribution, allowed for accurate explanation of data (Kulkarni & Sharma, 1998). The data (Kulkarni & Sharma, 1998), processed in variables $\varepsilon - \omega$ in Figure 5(b), show that burn rates for CMDB propellants change with ignition temperature either by the addition or absence of binding agent (AP). This binding agent corresponds to standard deviation parameter, of our developed model, employed for describing system's internal microstructure; this is evident from Figure 5(b). It is apparent that the present model developed with normal distribution of neighboring reaction cells is accurate for describing the combustion process of CMDB type of heterogeneous mixtures.

3.4. Ti–Si system
Here we correlate the standard deviation, parameter of developed model, with the stoichiometric coefficient of actual heterogeneous mixtures. The work (Rogachev & Baras, 2009) for combustion of Ti–Si system is considered, for comparison in the view of the developed model, since the amount of gas released is relatively small. The combustion process of Ti–Si mixture is represented as Ti + xSi, where x is referred as a stoichiometric coefficient. The plot for measured mean burn rate r and stoichiometric coefficient x is shown in Figure 3 from work (Rogachev & Baras, 2009), along with measured adiabatic temperature T_B. It is investigated that heterogeneous combustion of Ti + xSi mixtures occurs in range of x = [0.3, 1.5] as shown in work (Rogachev & Baras, 2009). Burning rate changes with change in stoichiometric coefficient. The maximum burning rate (38 mm/s) of mixture

coincides at highest value of burn temperature T_B. Burn rate is maximum at $x = 0.6$ which corresponds to synthesis $5Ti + 3Si \rightarrow Si_3Ti_5$.

Behavior of burn rate in range of $x = [1, 1.4]$ varies by previous trend: rate of burning decreases in the specified range even for the constant burning temperature. Such situation commonly arises in the combustion process of heterogeneous mixtures of $Ti + x \ Si$, where combustion process is associated with phase transformations and complex micro structural properties. Currently, no such alternate heterogeneous combustion models exist, for quantitative description of change in burn rate of thermite mixtures at constant burn temperature. Developed model, in view of normal distribution of neighboring reaction cells, has an additional scale of choice for tailoring neighboring reaction cells with variation of standard deviation parameter. Change of standard deviation parameter, for describing microstructure of system, results in change of burn rate, even at constant non-dimensional ignition temperature ε. Such a model explains change of burn rate for combustion of $Ti + x \ Si$ mixtures in the range of $x = [1, 1.4]$ where its burning temperature is constant as shown in Figure 3 from work (Rogachev & Baras, 2009). The correlation between standard deviation parameter and stoichiometry coefficient x is performed at burning temperature 1,840 k. From the developed model, it is observed that the burning rate decreases with an increase in the standard deviation parameter. It is sufficient to presume that standard deviation parameter depends on stoichiometry coefficient x. As described in earlier section, in the view of the developed model for combustion process it is not possible to systematize a stable combustion process for $\varepsilon > 0.42$ under several primary circumstances in the entire range of standard deviation parameter. $\varepsilon = 0.48$ is interpreted as natural combustion limit for the developed combustion model. Equivalent of 1,840 k is obtained, in nondimensional ignition temperature by Equation (27) from (Bharath et al., 2013), as $\varepsilon = 0.44$. The experimental burn rates at different stoichiometric mixtures for constant burning temperature are treated by:

$$\omega/\omega_{cr} = (x/x_{cr})(l_0/l_{0cr})$$

Figure 6(a) shows for Burn rate vs. stoichiometric coefficient and Burn rate vs. standard deviation σ shows the correlated values of stoichiometric coefficient and standard deviation. The solid line shows the numerical burn rates for different standard deviation parameters at $\varepsilon = 0.44$. Symbols represent experimental data. We obtain matched dependence of stoichiometric coefficient x and standard deviation σ for common value of burn rate. Now the dependence of standard deviation $\sigma(x)$ on stoichiometry coefficient x is obtained by performing best fit on experimental data of burn rate. For certainty, it is assumed $\sigma = 1$ for the lower combustion limits. Matched dependence $\sigma(x)$ is shown in the Figure 6(b). Thus, variations in standard deviation of neighboring reaction cells completely correlate dependence of burn rate on stoichiometry coefficient x even at constant burning temperatures. It is evident from Figure 6(b), analysis for the dependence of $\sigma(x)$, that the standard deviation parameter σ reaches a minimum value $\sigma \approx 0.01$ at $x = 0.6$.

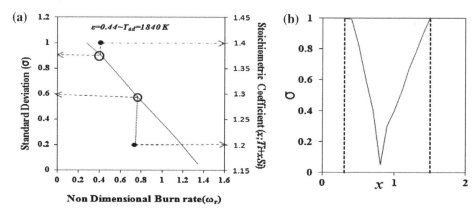

Figure 6. (a) Left shows comparison of experimental and theoretical nondimensional burn rates on stoichiometry x (Rogachev & Baras, 2009) and parameter σ. (b) Correlation between parameter σ and stoichiometry coefficient x; dashed lines represent inflammability limits.

As shown above, the less $\sigma \ll 1$ the more ordered is the system. Obtained dependence $\sigma(x)$ (Figure 6) shows that the Ti + xSi mixture becomes more disordered as it moves away from $x = 0.6$. Such an effect of microstructure of the system on combustion process is associated with binding of Ti and Si particles, mixing at different concentrations, over the combustor volume.

3.4.1. Modes of combustion front

The mechanism for propagation of combustion front, is of fundamental importance, and is analyzed with the role of internal microstructure. The developed model explains these combustible limits in the view of ignition time profiles (in the range [0.42, 0.48]), if we assume, that $\varepsilon > 0.48$ is not in the range of x = [0.3, 1.5]. Hence, our numerical calculations consider that $\varepsilon = 0.48$ for both lower combustible limits and this determined the value of T_{ign} = 990 K. The existence of ignition temperature in delay, for combustion of heterogeneous mixtures, initiates a limit to reaction front propagation even in the absence of heat losses and with uniform heat release. The instant at which the heat release of the active cell is adequate to increase the temperature of the particles to the predetermined ignition temperature is defined as combustion limit. We observe in our numerical calculations, as time is incremented the temperature of active cell reaches critical temperature which is still lower than the predetermined ignition temperature. This instant is determined as numerical combustion limit (Bharath et al., 2013; Dvoryankin et al., 1985; Rashkovskii, 2005; Rashkovskiy et al., 2010; Rogachev & Baras, 2009) of developed model. The numerical combustion limit for a periodic system is 0.5459 (Rashkovskii, 2005; Rashkovskiy et al., 2010), however for disordered system the combustion limit still further reduces and is observed to be 0.48. The combustion front cannot propagate through the system above the Combustion limit (Bharath et al., 2013; Dvoryankin et al., 1985; Rashkovskii, 2005; Rashkovskiy et al., 2010; Rogachev & Baras, 2009). Figure 7 shows the ignition time profiles of different normal disordered systems (σ = 0.05 and σ = 0.4) for higher ignition temperatures ranging from 0.42 to 0.48. The periodic system and disordered system with σ = 0.05 has similar characteristics for lower values of ignition temperature < 0.42, however as ignition temperature increases the microstructure of system plays a dominant role in deciding the combustion limit at higher values of ignition temperature. The combustion front propagates in quasi-homogeneous mode when the thickness of the adjacent particles is much smaller than the vicinity of reaction front.

The combustion front propagates in relay race mode when inter particle distance of the mixture is of the order of width of the vicinity of combustion front.

3.5. Thermite systems at higher combustion limits

The present section illustrates the comparison of experimental data for combustion of thermite systems with the theoretical model developed in earlier section. In due process, we try to establish the effect of distribution of heat release on combustion limit of thermite mixtures. The work of authors (Dvoryankin et al., 1985) investigate that the burning rate changes with burning

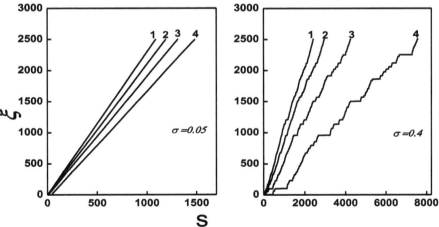

Figure 7. Higher ignition time profile for different Normal distributions.

Notes: 1 = 0.42(ε), 2 = 0.44(ε), 3 = 0.46(ε), 4 = 0.48(ε).

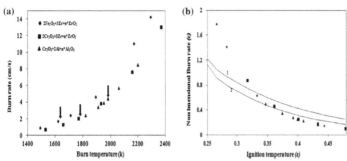

Figure 8. (a) Plot between burn rate and burn temperature for different thermite systems from work (Dvoryankin et al., 1985). (b) Comparison plot for theoretical and experimental dependencies $\omega(\varepsilon)$.

Notes: In (a) the arrow shows critical points at which combustion begins in oscillating mode. In (b) dots represent converted experimental data (Dvoryankin et al., 1985). Lines 1–2 are obtained from theoretical model developed for identical heat release, for normal distribution. Line 2 correspond to calculations for $\sigma = 0.98$ and line 1 correspond to (Rashkovskiy et al., 2010) γ $a = 0.7$.

temperature for a broad range by altering the inert diluter and percentage of inert diluter; in doing so the phenomenon of heat release in combustion process is not altered.

As established in above sections, by the developed model, the noticeable oscillations in those system commence at $\varepsilon_{cr(developed)} = 0.4$; we classify thermite mixtures from the work of the authors (Dvoryankin et al., 1985) based on their inflammability limits. Thus, we refer thermite systems from the work (Dvoryankin et al., 1985) into two: $\varepsilon_{cr(Work)} < \varepsilon_{cr(developed)}$, $\varepsilon_{cr(developed)} < \varepsilon_{cr(Work)}$. Here we consider thermite systems with $\varepsilon_{cr(developed)} < \varepsilon_{cr(Work)}$. Figure 10(a) comprise the data of work (Dvoryankin et al., 1985) ($2Fe_2O_3 + 3Zr + n*ZrO_2$, $2Cr_2O_3 + 3Zr + n*ZrO_2$, $Cr_2O_3 + 2Al + n*Al_2O_3$), processed in co-ordinates of burn rate and burning temperature. Neglecting heat losses and considering uniform heat release of reaction cells, the calculated burning temperature of the mixtures is identified as adiabatic temperature (T_{ad}) of the heterogeneous mixture. The average burn rate, during entire time of combustion process of different heterogeneous mixtures, determined in the work (Dvoryankin et al., 1985); corresponds to theoretical average burning rate Equation (6) (Figure 4). The arrows in Figure 8(a) correspond to commencement of oscillating mode for combustion process of different thermite mixtures. Treatment of experimental data (Dvoryankin et al., 1985) has been performed in nondimensional co-ordinates $\varepsilon - \omega$. Hence theoretical values of the critical parameters such as ε_{cr} and ω_{cr} are assumed as commencement of oscillating modes of combustion, practical in experiments. As established above, by the developed model, the noticeable oscillations in those system commence at $\varepsilon_{cr} = 0.42$; and is considered as critical parameters of developed model. Assuming the critical values of rate of burning (r_{cr}) and burning temperature ($T_{ad_{cr}}$), measured in experiments, corresponds to commencement of oscillating modes of combustion process for an actual system. Then Equation (27), from (Bharath et al., 2013), in accordance with Figure 4, critical non-dimensional burning rate $\omega_{cr} = \omega(\varepsilon_{cr})$ depends on the standard deviation of neighboring reaction cells; this implies that ω_{cr} depends on the system microstructure.

Experimental data (Dvoryankin et al., 1985) are treated using expressions (27)–(29) from Bharath et al. (2013), similar to method shown in Bharath et al. (2013), are shown in Figure 8(b). Line 2 (Figure 8(b)) shows the theoretical dependence for different values of $\sigma = const$. Line 1 shows the theoretical burn rates obtained for gamma distribution ($a = 0.7$) of neighboring reaction cells.

We assume that the change of non dimensional ignition temperature (ε) in diluting the system mixture is associated with change of the standard deviation parameter σ, which describe the system microstructure considering uniform heat release at all cells. Furthermore, we consider that combustion process of thermite mixtures from work Dvoryankin et al. (1985) can be explained by the respective dependence $\sigma(\varepsilon)$. The dependence $\sigma(\varepsilon)$ is coordinated for the condition of concurrence of theoretical dependence $\omega(\varepsilon)$ calculated by developed model with the experimental data (Dvoryankin et al., 1985) (Figure 8(b)). Such dependence $\sigma(\varepsilon)$ is shown in Figure 8(b). Theoretical dependence $\omega(\varepsilon)$, is calculated using expressions Equations (6), (5) in the view of matched dependence $\sigma(\varepsilon)$, is shown in Figure 8(b) (line 2); it describes experimental data (Dvoryankin et al., 1985) accurately for broad class of thermite mixtures. The value ω_{cr} = 0.25 was used in these calculations. Solid line 2 is the result of disordered system described with normal distribution of reaction cells and line 1 is from gamma distribution (a = 0.7). The Chi square at higher ignition temperatures (ε > 0.32) has been calculated for lines 1 and 2 with respect to experimental data; and we achieve a Chi square value of 0.054 for line 2 (normal distribution) and 0.523 for line 1 (gamma distribution a = 0.7). The data (Dvoryankin et al., 1985), processed in variables $\varepsilon - \omega$, show that combustion becomes impossible at ε = 0.49 ... 0.5, this reality correlates well with the theoretical results obtained by the developed model for uniform heat release at all reaction cells.

3.6. Disordered heat release

In the work (Dvoryankin et al., 1985), for different thermite mixtures of Fe_2O_3, it is observed that two thermite mixtures (Fe_2O_3 + 2Al + n*Al_2O_3, $2Fe_2O_3$ + 3Ti + n*TiO_2) have the inflammability limit less than $\varepsilon_{(work)}$< 0.4, and the steady-state mode of combustion is seen for further lower values of ε. The model developed in earlier sections with an assumption of uniform heat release (q_i) at all reaction cells could not explain for the thermite systems that have lower combustion limits. Currently, no such combustion models exist that can quantitatively explain the change of combustion limit for thermite mixtures. Unlike the above model, where the heat release is considered identical for all cells, the present section analyzes the effect of randomizing the heat release at all cells on combustion limit. It is established from the above developed model with normal distribution of neighboring reaction cells is more accurate with experimental data considered, here we consider normal distribution for positioning neighboring reaction cells with randomized heat release at each cell. We note that the system modeled with gamma distribution of neighboring reaction cells and randomizing the heat losess, the combustion front does not propagate even at lower ignition temperature (ε < 0.05). Hence the gamma distribution cannot be used in robust modeling of combustion process that considers randomizing heat release. Note the average heat release at each cells is maintained as unit. Heat release in the developed model is also viewed as the consumption of reactant. Randomizing the heat release not only allows us to study the combustion limit but also reveals the nature of distribution of heat releases of neighboring reaction cells.

An identical heat release at all point sources was assumed for the above numerical calculations. However, there can be a possibility of unequal heat release or unequal consumption of reactant (Gardner et al., 1999; Hwang et al., 1998; Mukasyan & Rogachev, 2008; Rogachev, 2003) at different cells. The heat release q_i is randomized and its effects on the combustion are studied in the present section. The heat release distribution, with two extreme cases has been considered. The system with small spread in the distribution of heat release is described by a normal distribution ($q_i \in$ [0.7–1.2]) and the more spread by a normal distribution ($q_i \in$ [0–7]), however the average heat release is maintained at one. The modeled systems are calculated numerically and analyzed for their effect on combustion process. We observe that the small spread of distribution of heat release does not show considerable effect on combustion process and the combustion process is same as with uniform heat release. However as the degree of randomizing the distribution of heat release increases we observe considerable change in the process of combustion of such systems. Figure 9 shows the ignition time profile for the microstructure of system described by σ = 0.05 and σ = 0.4 and with high disordered heat release.

It can be seen that the combustion process is affected by the disordered heat release. With the inclusion of the random heat release, the burn front moves slower and combustion stops are observed. The affects are more pronounced at higher ignition temperatures and higher disorder of the heat release. Figure 10 shows the comparison of the burn rate obtained for a system with high disorder in the heat release with identical heat release. As the randomness in heat release increases, the burn rates decreases with an increase in the ignition temperature. It can be observed that at higher ignition temperatures, the decrease in the burn rate is more when compared to an identical heat release. At any given ignition temperature, a system described by a low degree of randomness in the position of its cells and higher degree of disorderness in nature of heat releases shows a higher

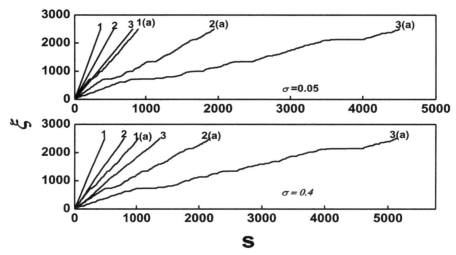

Figure 9. Comparison of Ignition time profiles for Identical heat release (1, 2, 3) and disordered heat release (1(a), 2(a), 3(a)) (q_i). 1 = 0.15(ε), 2 = 0.25(ε), 3 = 0.35(ε).

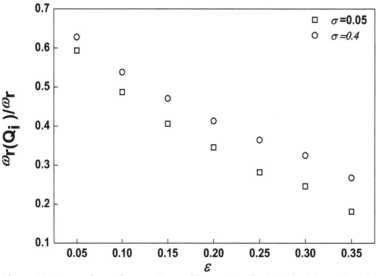

Figure 10. Comparison of percentage of burn rates for identical (boxes) and disordered (circles) heat release.

decrease in the burn rate. The combustion limit for the present disordered system with randomizing heat release is still further reduced to 0.40 from 0.48. And we detect oscillations of ignition time profiles at ε = 0.35. Our numerical calculations show that the combustion process of a system is affected by randomizing the heat release that leads to slow heating of system. Thus, the distribution of heat release in the system affects the combustion process (Hwang et al., 1998; Mukasyan & Rogachev, 2008; Rogachev, 2003) and hence cannot be neglected while modeling the combustible system.

As the clusterization of heat release increases then the combustion process differs to a great extent and this increases with an increase in the ignition temperature. Heat release study would be helpful for robust modeling of combustible system which not only accounts for microstructure of system but also the different possibilities of heat releases.

3.6.1. Thermite systems with inert diluents
Present section compares the experimental data for thermite mixtures (Dvoryankin et al., 1985), that consist of powder components (Fe$_2$O$_3$ + 2Al + n*Al$_2$O$_3$, 2Fe$_2$O$_3$ + 3Ti + n*TiO$_2$) mixed with inert diluter, with theoretical model developed in section 3.2 where the heat release is disordered. Powdered

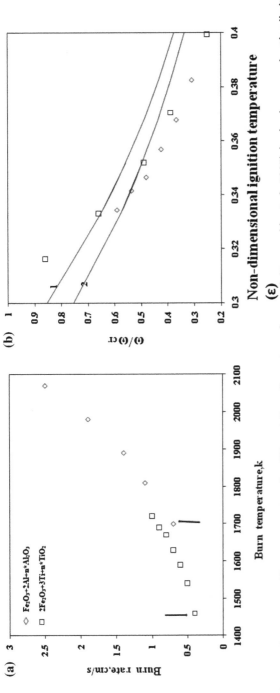

Figure 11. (a) Plot between burn rate and burn temperature for different thermite systems from work (Dvoryankin et al., 1985), having lower combustion limits, based on data of work (Dvoryankin et al., 1985). (b) Comparison plot (enlarged scale) between theoretical and experimental dependencies $\omega(\varepsilon)$. Symbols are the converted experimental data (Dvoryankin et al., 1985).

Notes: In (a) the arrow shows critical points at which combustion begins in oscillating modes. In (b) solid lines are the theoretical dependence for randomized heat release and normal distribution with microstructure described by line 1 $\sigma = 0.05$ and by line 2 $\sigma = 0.40$.

components are capable of exothermic transformation. The temperature for burning the heterogeneous mixtures and their rate of burning can be altered for a broad range by changing the diluter and also by the amount of inert diluter; in doing so the lower inflammability limit in combustion is changed. The values of burning rate and burning temperature of heterogeneous mixture that contain different inert diluters and different percentages of inert diluter are collected from the work (Dvoryankin et al., 1985). Neglecting heat losses and considering randomizing distribution of heat release, the burning temperature of the heterogeneous mixtures is identified as adiabatic temperature (T_{ad}) of the mixture. It is established above, from the developed model (considering disordered heat releases), that the detectable oscillations in such mixtures begin at $\varepsilon_{cr} = 0.35$; we classify the thermite systems from the work of the authors (Dvoryankin et al., 1985) based on their combustion limits. Present section refers to thermite mixtures from the work (Dvoryankin et al., 1985) for which $\varepsilon_{cr(Work)} < \varepsilon_{cr(developed)}$. Figure 11(a) contains the data of work (Dvoryankin et al., 1985) ($Fe_2O_3 + 2Al + n*Al_2O_3$, $2Fe_2O_3 + 3Ti + n*TiO_2$), performed in the co-ordinates of burn rate and burning temperature. The arrows in Figure 11(a) correspond to the beginning of oscillating (critical) modes of combustion of thermite mixtures. Treatment of experimental data (Dvoryankin et al., 1985) is performed, similar to method in Bharath et al. (2013), in non dimensional co-ordinates of $\varepsilon - \omega$. Hence it is considered that commencement of oscillating mode of combustion process, noticeable during experiments, is associated with critical values of theoretical parameters such as ε_{cr} and ω_{cr}. As established above the noticeable oscillations of such heterogeneous mixture commence at $\varepsilon_{cr} = 0.35$, this parameter is considered as critical value of inflammability limit of the developed model. Assuming critical values of rate of burning r_{cr} and burning temperature $T_{ad_{cr}}$, noticeable during experiments, corresponds to commencement of oscillating modes of combustion in an actual system. Experimental data (Dvoryankin et al., 1985) now converted using Equations (27–29) from (Bharath et al., 2013) are shown in Figure 11(b). Lines 1–2 (Figure 11(b)) show for different values $\sigma = const$ and randomized q_i.

We speculate the changing of nondimensional ignition temperature (ε) in dilution of thermite mixtures is associated with change of standard deviation (σ), which describe the system internal structure. And also assume change of combustion limit is associated with randomizing of distribution of heat release. Moreover, we consider the thermite systems and their combustion limit can be described by the dependence q_i and $\sigma(\varepsilon)$. The value of parameter q_i and $\sigma(\varepsilon)$ is matched from the concurrence of numerical (developed model) burn rates $\omega(\varepsilon)$ with the experimental data (Dvoryankin et al., 1985) (Figure 11(b)). Such dependence of $\sigma(\varepsilon)$ and q_i is shown in Figure 11(b) (line 2). Theoretical dependence $\omega(\varepsilon)$, in view of the matched dependence of parameters $\sigma(\varepsilon)$ and q_i, is shown in Figure 11(b) (line 2); it appropriately explains the experimental data (Dvoryankin et al., 1985) of thermite mixtures having lower inflammability limits. The value $\omega_{cr} = 0.26$ was used for above calculations. Experimental data (Dvoryankin et al., 1985) allow estimating an inflammability limit of ignition temperature (ε) for each mixture, beyond which a stable combustion is not possible. The experimental data (Dvoryankin et al., 1985), now converted in the coordinates of $\varepsilon - \omega$, show that stable combustion is not possible at values $\varepsilon = 0.35 \ldots 0.4$; this reality agrees well with numerical results of the developed model obtained by randomizing distribution of heat release.

4. Conclusions

Gasless heterogeneous combustion system is modeled as a one-dimensional chain of combustible particles. The positions of adjacent reaction cells with significance to classification of distribution are described by normal, uniform, and beta. The reactions at higher ignition temperatures become complex and in such situations minute changes in internal microstructure can play a crucial role in affecting combustion process. In such situations, normal distribution of neighboring reaction cells has an additional scale of choice characterized by the parameter known as standard deviation. Relay race condition of combustion reaction front is observed for higher values of ignition temperature and higher degree of randomness in microstructure of system where as quasi homogeneous condition of combustion reaction front is observed for lower values of ignition temperature. Burn rates of a normal distributed disordered system are observed to be in the range of burn rates for periodic to most randomly distributed system. Generalized theoretical expression for the dependence of normalized burn rates to periodic burn rates on variance of the distribution is obtained. The ability to introduce randomness through normal distribution of adjacent reaction cells allows for description of results for a wide range of experiments such as CMDB propellants, Ti + xSi mixtures, and thermite systems. Experimental data of pyrotechnic mixtures are closely explained by different distributions with the priority of Normal, Gamma.

Randomizing the nature of distribution of heat release explains combustion process for thermite systems with lower inflammability limits. The theoretical combustible limit for disordered heat release in system is observed to be 0.38 and correlates with experimental inflammability limit, which is much less than that of identical heat release with normal distribution of active adjacent cells. We understand that a one-dimensional description of disperse system cannot reveal the flame structure; however, detailed numerical study on effect of microstructure, with significance to classification of random distributions, on combustion process shall list the priorities of distribution. The information from this work would be referred for multi-dimensional modeling of heterogeneous combustion process.

Acknowledgments
Authors acknowledge DRDO for funding and CMSD for providing computation facility.

Funding
The authors received funding from Defence Research and Development Organisation (DRDO), India.

Author details
Tarun Bharath Naine[1]
E-mail: nainetarun@gmail.com
Manoj Kumar Gundawar[1]
E-mails: manojsp@uohyd.ernet.in, manoj@uohyd.ac.in
ORCID ID: http://orcid.org/0000-0002-4016-0354
[1] Advanced Centre of Research in High Energy Materials (ACRHEM), University of Hyderabad, Hyderabad 500046, Telangana, India.

References
Beckstead, M. W., & McCarty, K. P. (1982). Modeling calculations for HMX composite propellants. *The American Institute of Aeronautics and Astronautics Journal, 20*, 106–115.

Bharath, N. T., Rashkovskiy, S. A., Tewari, S. P., & Gundawar, M. K. (2013). Dynamical and statistical behavior of discrete combustion waves: A theoretical and numerical study. *Physical Review E, 87*, 042804. http://dx.doi.org/10.1103/PhysRevE.87.042804

Candel, Sebastien (2002). Combustion dynamics and control: Progress and challenges. *Proceedings of the Combustion Institute, 29*, 1–28. http://dx.doi.org/10.1016/S1540-7489(02)80007-4

Denisyuk, A. P., Shabalin, V. S., & Shepelev, Y. G. (1998). Combustion of condensed systems consisting of HMX and a binder capable of self-sustained combustion. *Combustion, Explosion and Shock Waves, 34*, 534–542.

Dvoryankin, A. V., Strunina, A. G., & Merzhanov, A. G. (1985). Stability of combustion in thermite systems. *Combustion, Explosion, and Shock Waves, 21*, 421. http://dx.doi.org/10.1007/BF01463412

Favier, C. (2004). Percolation model of fire dynamic. *Physics Letters A, 330*, 396–401. http://dx.doi.org/10.1016/j.physleta.2004.07.053

Gardner, R. H., Romme, W. H., & Turner, M. G. (1999). Predicting forest fire effects at landscape scales. In D. J., Mladenoff & W. L., Baker (Eds.), *Spatial modeling of forest landscapes: Approaches and applications* (pp. 163–185). Cambridge: Cambridge University Press.

Goroshin, S., Lee, J. H. S., & Shoshin, Y. (1998). Effect of the discrete nature of heat sources on flame propagation in particulate suspensions. *Symposium (International) on Combustion, 27*, 743–749. http://dx.doi.org/10.1016/S0082-0784(98)80468-2

Gross, M. (2010). *Towards a predictive propellant burning rate model based on high-fidelity numerical calculations.* 46th AIAA/ASME/SAE/ASEE Joint Propulsion Conference & Exhibit, Joint Propulsion Conferences. Nashville, TN. doi:10.2514/6.2010-6914

Humphrey, M. F. (1971). Solid propellant burning-rate modification. (JPL Quarterly Technical Review) *Propulsion division, 1.*

Hwang, S., Mukasyan, A. S., & Varma, A. (1998). Mechanisms of combustion wave propagation in heterogeneous reaction systems. *Combustion and Flame, 115*, 354–363. http://dx.doi.org/10.1016/S0010-2180(98)00016-9

Kerstein, A. R. (1987). Percolation model of polydisperse composite solid propellant combustion. *Combustion and Flame, 69*, 95–112. http://dx.doi.org/10.1016/0010-2180(87)90023-X

Kubota, N. (1978). Role of additives in combustion waves and effect on stable combustion limit of double-base propellants. *Propellants, Explosives, Pyrotechnics, 3*, 163–168. http://dx.doi.org/10.1002/(ISSN)1521-4087

Kubota, N. (2002). *Propellants and explosives: Thermochemical aspects of combustion.* Willey-VCH Verlag GmbH and Co. KGaA.

Kubota, N., & Okuhara, H. (1989). Burning rate temperature sensitivity of HMX propellants. *Journal of Propulsion and Power, 5*, 406–410. doi:10.2514/3.23169

Kulkarni, A. R., & Sharma, K. C. (1998). Burn rate modelling of solid rocket propellants. *Defence Science Journal, 48*, 119–123. http://dx.doi.org/10.14429/dsj.48.3876

Mukasyan, A. S., & Rogachev, A. S. (2008). Discrete reaction waves: Gasless combustion of solid powder mixtures. *Progress in Energy and Combustion Science, 34*, 377–416. http://dx.doi.org/10.1016/j.pecs.2007.09.002

Rashkovskii, S. A. (1999). Structure of heterogeneous condensed mixtures. *Combustion, Explosion, and Shock Waves, 35*, 523–531. http://dx.doi.org/10.1007/BF02674497

Rashkovskii, S. A. (2005). Hot-spot combustion of heterogeneous condensed mixtures. Thermal percolation. *Combustion, Explosion and Shock Waves, 41,* 35–46. http://dx.doi.org/10.1007/s10573-005-0004-4

Rashkovskiy, S. A., Kumar, G. M., & Tewari, S. P. (2010). One-dimensional discrete combustion waves in periodical and random systems. *Combustion Science and Technology, 182,* 1009–1028. http://dx.doi.org/10.1080/00102200903544263

Rogachev, A. S. (2003). Microheterogeneous mechanism of gasless combustion. *Combustion, Explosion, and Shock Waves, 39,* 150–158. http://dx.doi.org/10.1023/A:1022956915794

Rogachev, A. S., & Baras, F. (2007). Models of SHS: An overview. *International Journal of Self-Propagating High-Temperature Synthesis, 16,* 141–153. http://dx.doi.org/10.3103/S1061386207030077

Rogachev, A. S., & Baras, F. (2009). Dynamical and statistical properties of high-temperature self-propagating fronts: An experimental study. *Physical Review E, 79,* 026214. http://dx.doi.org/10.1103/PhysRevE.79.026214

Varma, A., & Mukasyan, A. S. (2001). Dynamics of self-propagating reactions in heterogeneous media: Experiments and model. *Chemical Engineering Science, 56,* 1459–1466. http://dx.doi.org/10.1016/S0009-2509(00)00371-7

Viegas, D. G. (1998). *Philosophical Transactions of the Royal Society A: Mathematical, Physical and Engineering Sciences, 356,* 2907–2928. doi:10.1098/rsta.1998.0303

Willcox, M. A., Brewster, M. Q., Tang, K. C., & Stewart, D. S. (2007). Solid propellant grain design and burnback simulation using a minimum distance function. *Journal of Propulsion and Power, 23,* 465–475. doi:10.2514/1.22937

8

Optimization of thermal performance of a smooth flat-plate solar air heater using teaching–learning-based optimization algorithm

R. Venkata Rao[1] and Gajanan Waghmare[1]*

*Corresponding author: Gajanan Waghmare, Department of Mechanical Engineering, S.V. National Institute of Technology, Ichchanath, Surat, Gujarat 395 007, India
E-mail: waghmaregaju@yahoo.com
Reviewing editor: Duc Pham, University of Birmingham, UK

Abstract: This paper presents the performance of teaching–learning-based optimization (TLBO) algorithm to obtain the optimum set of design and operating parameters for a smooth flat plate solar air heater (SFPSAH). The TLBO algorithm is a recently proposed population-based algorithm, which simulates the teaching–learning process of the classroom. Maximization of thermal efficiency is considered as an objective function for the thermal performance of SFPSAH. The number of glass plates, irradiance, and the Reynolds number are considered as the design parameters and wind velocity, tilt angle, ambient temperature, and emissivity of the plate are considered as the operating parameters to obtain the thermal performance of the SFPSAH using the TLBO algorithm. The computational results have shown that the TLBO algorithm is better or competitive to other optimization algorithms recently reported in the literature for the considered problem.

Subjects: Engineering Mathematics; Renewable Energy; Energy and Fuels

Keywords: smooth flat plate solar air heater; thermal efficiency; teaching–learning-based optimization algorithm

ABOUT THE AUTHOR

R. Venkata Rao is a professor in the Department of Mechanical Engineering of S.V. National Institute of Technology, Surat, Gujarat (India). He received his BTech degree from Nagarjuna University, his M.Tech degree from BHU, Varanasi, and his PhD degree from BITS, Pilani, India. He has about 23 years of teaching and research experience. He has authored about 260 research papers published in various reputed international journals and conference proceedings. He is also on the editorial boards of various international journals. His research interests include: advanced engineering optimization techniques and the applications, fuzzy multiple attribute decision-making, advanced manufacturing technology, automation and robotics.

PUBLIC INTEREST STATEMENT

Solar air heater is extensively used nowadays in various applications like space heating, seasoning of timber, curing of industrial products, etc. It consists of an absorbed plate with parallel plate below forming a small passage through which air is to be heated and flow. But the main drawback of solar air heater is low thermal efficiency, since the heat transfer coefficient is less between the absorber plate and air. Sometimes surfaces are roughed or longitudinal fines are used or geometrical parameters are changed to increase the thermal efficiency. In the present work, optimal set of operating parameters are investigated using the teaching–learning-based optimization algorithm, at which the thermal performance of a smooth flat plate solar air heater could be maximum. The computational results have shown that the TLBO algorithm is better or competitive to other optimization algorithms recently reported in the literature for the considered problem.

1. Introduction

Solar air heating is a solar thermal technology, in which the energy from sun is captured by an absorbing medium and used to heat air. Solar air heating is extensively used nowadays in commercial and industrial applications. Solar air heaters are simple in design and construction, but efficiency of flat plate solar air heater is low because of low convective heat transfer coefficient between the absorber plate and the flowing air that increases absorber plate temperature, leading to higher heat losses to environment. Low value of heat transfer coefficient is due to presence of laminar sublayer that can be broken by providing artificial roughness on heat transferring surface.

The use of flat-plate solar collectors to heat air to relatively low temperature has become a common practice in numerous applications from space heating to food dehydration industry (Hegazy, 1996). Hegazy (1996) optimized the flow channel depth for a conventional flat-plate solar air heater. The author had derived an expression for estimating the channel depth-to-length ratio that yields an outlet air temperature equal to the absorber plate mean temperature in terms of flow pumping power. This expression is of great importance for designers of this type of solar air heater. A parametric study was also carried out to investigate the effect of the channel depth on collector useful heat gain of collector over a wide range of D/L (depth of flow channel/length of absorber plate) ratios, and for different pumping power requirements.

Gupta, Solanki, and Saini (1997) explained that the systems operating in a specified range of Reynolds number show better thermohydraulic performance depending upon the insulation. Ammari (2003) presented a mathematical model for computing the thermal performance of a single-pass flat-plate solar air collector. The author had investigated the influence of the addition of the metal slats on the efficiency of solar collector with the help of the model developed. The effect of volume air flow rate, collector length, and spacing between the absorber and bottom plates on the thermal performance of the solar air heater was investigated.

Mittal and Varshney (2006) investigated thermohydraulic performance of a wire mesh-packed solar air heater having its duct packed with blackened wire screen matrices of different geometrical parameters (wire diameter and pitch). The authors had concluded that the Reynolds number was a strong parameter affecting the effective efficiency. Also, it was found that for higher values of the temperature rise, the effective efficiency values closely followed the thermal efficiency values, whereas there was an appreciable difference in the lower range of temperature rise values. The authors had also commented that merely the porosity of the bed does not govern the performance. Kalogirou (2006) used artificial neural networks (ANN) for the prediction of the performance parameters of flat-plate solar collectors. Six ANN models were developed for the prediction of the standard performance of collectors.

Layek, Saini, and Solanki (2007) optimized solar air heater having chamfered rib-groove roughness on absorber plate. The entropy generation in the duct of solar air heater having repeated transverse chamfered rib-groove roughness on one broad wall was studied numerically. The authors had concluded that the roughness parameters like relative roughness pitch, relative roughness height, relative groove position, chamfer angle, and flow Reynolds number had a combined effect on the heat transfer as well as fluid friction.

Improving the thermal performance by enhancing the heat transfer rate and reducing friction losses depends on the geometrical parameters of the solar air heater and hence, there is a need for optimization of design and operating parameters of the solar air heater. Varun and Siddhartha (2010) used genetic algorithm to investigate the thermal performance optimization of a flat-plate solar air heater. The authors considered different systems and operating parameters to obtain maximum thermal performance. Thermal performance was obtained for different Reynolds numbers, emissivity of the plate, tilt angle, and number of glass plate by using genetic algorithm.

Varun, Sharma, Bhat, and Grover (2011) implemented a stochastic iterative perturbation technique to obtain the optimum set of different system and operating parameters, such as the number of glass cover plate, emissivity of the plate, mean plate temperature, rise in temperature, tilt angle, and solar radiation intensity for different Reynolds numbers. El-Sebaii, Aboul-Enein, Ramadan, Shalaby, and Moharram (2011) presented an analytical model for the air heater with flat and V-corrugated plates. The authors had investigated the thermal performance of double-pass flat- and V-corrugated-plate solar air heaters theoretically and experimentally. The effect of mass flow rate of air on pressure drop, thermal and thermohydraulic efficiencies of the flat- and V-corrugated-plate solar air heater were also investigated.

Lanjewar, Bhagoria, and Sarviya (2011) presented experimental investigation of heat transfer and friction factor characteristics of a rectangular duct roughened with W-shaped ribs on its underside on one broad wall arranged at an inclination with respect to flow direction. The authors had compared the results of heat transfer and friction factor with those for a smooth duct under similar flow and thermal boundary conditions to determine the thermohydraulic performance. Correlations were also developed for heat transfer coefficient and friction factor for the roughened duct.

Tanda (2011) discussed the performance of solar air heater ducts with different types of ribs on the absorber plate. All the rib-roughened channels performed better than the reference smooth channel in the medium–low range of the investigated Reynolds number values. Gill, Singh, and Singh (2012) designed, fabricated, and tested two low-cost solar air heaters, i.e. single glazed and double glazed. The collector efficiency factor, heat-removal factor based on air outlet temperature, and air inlet temperature for solar air heaters were also determined.

There is an increasing interest among researchers in the design, development, and optimization of a smooth flat-plate solar air heater (SFPSAH) over past few decades. Siddhartha, Sharma, and Varun (2012) used particle swarm optimization algorithm for optimization of thermal performance of SFPSAH. The authors had carried out simulation for three different cases using the climatic condition data of Hamirpur city of India to investigate the thermal performance of SFPSAH. Maximization of thermal efficiency was set as an objective function. Siddhartha, Chauhan, Varun, and Sharma (2012) used simulated annealing algorithm to optimize the thermal performance of SFPSAH and predicted the optimum set of design and operating parameters.

Chamoli, Chauhan, Thakur, and Saini (2012) presented an extensive study of the research carried out on double-pass solar air heater. Karwa and Chitoshiya (2013) presented an experimental study of thermohydraulic performance of a solar air heater with 60° V-down discrete rib roughened on the airflow side of the absorber plate along with a smooth duct air heater. The authors claimed that the thermal efficiency was increased by 12.5–20% due to the roughness on the absorber plate depending on the airflow rate; higher enhancement was at the lower flow rate.

It has been observed that only few researches had attempted the optimization of flat-plate solar air heater by considering the different system and operating parameters to obtain maximum thermal performance (Varun & Siddhartha, 2010; Siddhartha, Chauhan, et al., 2012). Varun and Siddhartha (2010) used GA, Siddhartha, Chauhan, et al. (2012) used PSO and Siddhartha, Sharma, et al. (2012) used simulated annealing (SA) for optimization of thermal performance of a SFPSAH. However, the parameter setting of the GA, PSO, and SA algorithms is a serious problem which influences their efficiency and affects the performance of the algorithms, for example, GA requires the crossover probability, mutation rate, and selection method; PSO requires learning factors, the variation of weight, and the maximum value of velocity; SA requires temperature decrement. Similarly, the other advanced optimization algorithms like artificial bee colony (ABC) requires number of employed bees, onlooker bees, and value of limit; harmony search (HS) requires the harmony memory consideration rate, pitch adjusting rate, and number of improvisations. Unlike other optimization techniques, a recently developed optimization technique, namely teaching–learning-based optimization (TLBO) algorithm does not require any algorithm parameters to be tuned, thus making the implementation of TLBO algorithm simpler. This algorithm requires only the common control parameters and does not require any algorithm-specific control parameters. In the literature, it is observed that the TLBO algorithm is not yet used in the field of optimization of a SFPSAH. Hence, in this paper, TLBO algorithm is used to estimate the optimal performance of a SFPSAH, with various effective parameters. The next section presents the details of the TLBO algorithm.

2. TLBO algorithm

TLBO is a teaching–learning process-inspired algorithm proposed by Rao, Savsani, and Vakharia (2011a, 2011b) and Rao, Savsani, and Balic (2011) based on the effect of influence of a teacher on the output of learners in a class. The algorithm describes two basic modes of the learning: (1) through teacher (known as teacher phase) and (2) interacting with the other learners (known as learner phase). In this optimization algorithm, a group of learners is considered as population and different subjects offered to the learners are considered as different design variables of the optimization problem, and a learner's result is analogous to the "fitness" value of the optimization problem. The best solution in the entire population is considered as the teacher. The design variables are actually the parameters involved in the objective function of the given optimization problem and the best solution is the best value of the objective function. The working of TLBO is divided into two parts, "Teacher phase" and "Learner phase". The flow chart of TLBO algorithm is shown in Figure 1. Working of both these phases is explained below.

2.1. Teacher phase

It is the first part of the algorithm where learners learn through the teacher. During this phase, a teacher tries to increase the mean result of the class in the subject taught by him or her depending on his or her capability. At any iteration i, assume that there are "m" number of subjects (i.e. design variables), "n" number of learners (i.e. population size, $k = 1, 2, ... , n$), and $M_{j,i}$ be the mean result of the learners in a particular subject "j" ($j = 1, 2, ... ,m$) The best overall result $X_{total-kbest, i}$ considering all the subjects together obtained in the entire population of learners can be considered as the result of best learner $kbest$. However, as the teacher is usually considered as a highly learned person who trains learners, so that they can have better results, the best learner identified is considered by the algorithm as the teacher. The difference between the existing mean result of each subject and the corresponding result of the teacher for each subject is given by,

$$\text{Difference\_Mean}_{j,k,i} = r_i \left(X_{j,kbest,i} - T_F \cdot M_{j,i} \right) \tag{1}$$

where, $X_{j, kbest, i}$ is the result of the best learner (i.e. teacher) in subject j. T_F is the teaching factor. It is important to note here that after conducting a number of computational experiments on various benchmark functions (Rao & Waghmare, 2014), it was observed that the best value of objective function can be achieved when T_F value is taken as 1. So, in the TLBO algorithm, the value of T_F is considered as 1 and the term "T_F" is removed in the TLBO algorithm. Hence, Equation 1 is rewritten as:

$$\text{Difference\_Mean}_{j,k,i} = r_i \left(X_{j,kbest,i} - M_{j,i} \right) \tag{2}$$

Based on the Difference_Mean$_{j, k, i}$, the existing solution is updated in the teacher phase according to the following expression.

$$X'_{j,k,i} = X_{j,k,i} + \text{Difference\_Mean}_{j,k,I} \tag{3}$$

where $X'_{j,k,i}$ is the updated value of $X_{j, k, i}$. $X'_{j,k,i}$ is accepted if it gives better function value. All the accepted function values at the end of the teacher phase are maintained and these values become the input to the learner phase. The learner phase depends upon the teacher phase.

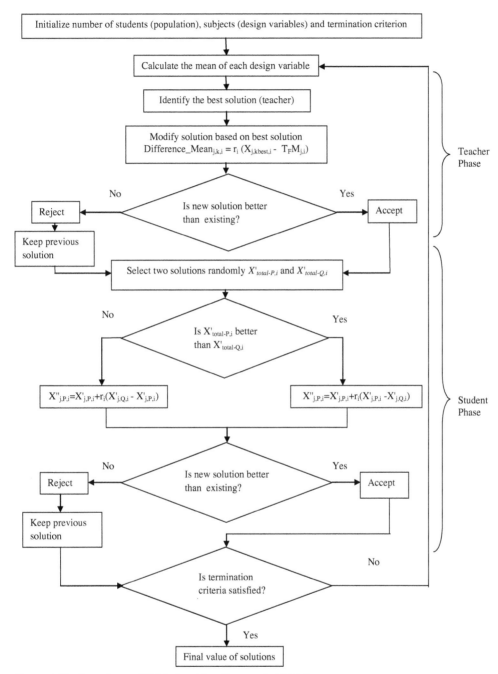

Figure 1. The flow chart of TLBO algorithm (Rao et al., 2011a).

2.2. Learner phase

It is the second part of the algorithm where learners increase their knowledge by interaction among themselves. A learner interacts randomly with other learners for enhancing his or her knowledge. A learner learns new things if the other learner has more knowledge than him or her. Considering a population size of "n", the learning phenomenon of this phase is expressed below:

Randomly select two learners P and Q, such that $X'_{total-P,i} \neq X'_{total-Q,i}$ (where $X'_{total-P,i}$ and $X'_{total-Q,i}$ are the updated values of $X_{total-P,i}$ and $X_{total-Q,i}$, respectively, at the end of teacher phase)

$$X''_{j,P,i} = X'_{j,P,i} + r_i \left(X'_{j,P,i} - X'_{j,Q,i} \right), \quad \text{If } X'_{total-P,i} < X'_{total-Q,i} \tag{4a}$$

$$X''_{j,P,i} = X'_{j,P,i} + r_i \left(X'_{j,Q,i} - X'_{j,P,i} \right), \quad \text{If } X'_{total-Q,I} < X'_{total-P,i} \tag{4b}$$

$X''_{j,P,i}$ is accepted if it gives a better function value.

   The TLBO algorithm has been already tested on several constrained and unconstrained benchmark functions and proved better than the other advanced optimization techniques (Rao & Patel, 2012, 2013; Rao & Waghmare, 2014) had evaluated the performance of the TLBO algorithm over a set of multi-objective unconstrained and constrained test functions and the results were compared against the other optimization algorithms. The TLBO algorithm was observed to outperform the other optimization algorithms for the multi-objective unconstrained and constrained benchmark problems.

   Waghmare (2013) presented the correct understanding about the TLBO algorithm in an objective manner and comments were made on the note of Črepinšek, Liu, and Mernik (2012). Yu, Wang, and Wang (2014) used improved TLBO for numerical and engineering optimization problems. The authors mentioned that the claim made by Črepinšek, Liu, and Mernik (2014) that Waghmare (2013) used different success rates was unsuitable. The comparisons of evolutionary algorithms conducted by Veček, Mernik, and Črepinšek (2014) attempted to cast the TLBO algorithm in a poor light, although this attempt may also be seen as not meaningful. The findings were simply comparisons of the basic TLBO algorithm with different modified versions of DE and did not consider other important algorithms, such as the GA, SA, PSO, and ACO (Yu et al., 2014). It may be mentioned that various researchers like Niknam, Azizipanah-Abarghooee, and Rasoul Narimani (2012), Rao, Kalyankar, and Waghmare (2014), Baykasoğlu, Hamzadayi, and Köse (2014), Satapathy and Naik (2014), Medina, Das, and Coello (2014), Basu (2014), Zou, Wang, Hei, Chen, and Yang (2014), Camp and Farshchin (2014), Moghadam and Seifi (2014) and Sultana and Roy (2014) proved the better performance of the TLBO algorithm as compared to the other evolutionary algorithms. Hence, the TLBO algorithm is attempted in the present work for the optimization of thermal performance of SFPSAH.

3. Thermal performance of solar air heater
The thermal performance of SFPSAH is investigated using the TLBO algorithm based on heat transfer phenomena (ASHRAE Standards) and calculation of flat-plate collector loss coefficients (Klein, 1975).

   Figure 2 presents the zenith angle, angle of incidence, tilts angle, and azimuth angle for a tilted surface (Twidell & Weir, 2005). The angle between the sun direction and the normal direction of a tilted surface can be represented as:

$$\cos \theta = \cos \theta_z \cos \beta + \sin \theta_z \sin \beta \cos(\gamma_s - \gamma) \tag{5}$$

where θ is the angle of incidence, θ_z is the solar zenith, β is the tilt angle, γ_s is the azimuth angles, and γ is the azimuth angle for a tilted surface.

   The design parameters are: number of glass cover plates, irradiance, and Reynolds number and the operating parameters are: wind velocity, plate tilt angle, emissivity of the plate, and ambient temperature. The thermal performance of a SFPASH can be predicted on the basis of detailed considerations of heat transfer processes and correlations for heat transfer coefficient, heat removal factor, etc. The objective function for thermal performance of SFPSAH can be proposed as given by ASHRAE Standards (1997) and expressed by the following equation:

$$\text{Maximize } \eta_{th} = F_0 \left[\tau\alpha - \left(\frac{T_0 - T_i}{S} \right) U_0 \right] \tag{6}$$

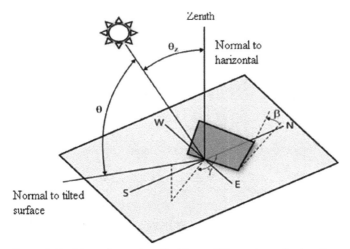

Figure 2. Zenith angle, angle of incidence, tilts angle, and azimuth angle for a tilted surface (Twidell & Weir, 2005).

The top loss coefficient is determined using relation (7) (Klein, 1975).

The different relations used for calculating overall loss coefficient (U_0), heat removal factor at outlet (F_0), and temperature rise ($T_0 - T_i$) are computed using relation (7), (11) and (13), respectively.

$$U_0 = \left[\frac{N}{\left(C/t_p \right) \left[(t_p - t_a)/(N+f') \right]^e} + \frac{1}{h_w} \right]^{-1}$$

$$+ \frac{\sigma(T_p + T_a)(T_p^2 + T_a^2)}{\left[\epsilon_p + 0.00591Nh_w \right]^{-1} + \left[(2N + f' - 1 + -0.133\epsilon_p)/\epsilon_p \right] - N} + \frac{k_i}{t} \qquad (7)$$

where

$$f' = \left(1 + 0.089h_w - 0.11h_w\epsilon_p \right)(1 + 0.07866N) \qquad (8)$$

$$C = 520 \left(1 - 0.000051\beta^2 \right) \qquad (9)$$

$$e = 0.43 \left(1 - 100T_p^{-1} \right) \qquad (10)$$

Heat removal factor at outlet (F_0) can be expressed as:

$$F_0 = \frac{Gc_p}{U_0} \left[1 - \exp\left(\frac{-U_0 F'}{GC_p} \right) \right] \qquad (11)$$

where

$$F' = \frac{\left[0.024R_e^{0.8}P_r^{0.4}\frac{\lambda}{d} \right]}{\left[0.024R_e^{0.8}P_r^{0.4}\frac{\lambda}{d} + U_0 \right]} \qquad (12)$$

The temperature rise $(T_0 - T_i)$ is computed by the following equation.

$$(T_0 - T_i) = \left[\frac{\left\{ (\tau\sigma)S - U_0(T_p - T_a) \right\}}{mc_p} \right] \cdot A_c \qquad (13)$$

The constraints of the problem are:

- $1 \le N \le 3$; N is varied in steps of 1.
- $600 \le S \le 1{,}000$; S is varied in steps of 200.
- $2{,}000 \le Re \le 20{,}000$; Re is varied in steps of 2,000.

The computations were carried out using TLBO algorithm for three different cases using the climatic condition data of the city of Hamirpur, India, situated between 31°25′–31°52′ N (latitude) and 76°18′–76°44′ E (longitude).

The other climatic conditions of the city of Hamirpur, India, are as follows:

$$1 \le v \le 3$$

$$280 \le T_a \le 310$$

where v is the wind velocity (m/s) and T_a is ambient temperature.

The following three different cases are considered (Siddhartha, Chauhan, et al., 2012).

Case 1: Obtain the value of V and T_a through TLBO algorithm and generate ϵ_p (0.85–0.95) and β (0°–70°) randomly.

Case 2: Obtain the value of β and T_a through TLBO algorithm and generate ϵ_p (0.85–0.95) and v (1–3) randomly.

Case 3: Obtain the value of V and β through TLBO algorithm and generate ϵ_p (0.85–0.95) and T_a (280–310 K) randomly for a fixed value of N (1, 2 and 3) and fixed S (600, 800, and 1,000 W/m^2) and varying Re ranging from 2,000 to 20,000 in an incremental step of 2,000. The next section explains the detailed results and discussion.

4. Results and discussion

To check the effectiveness of the TLBO algorithm, extensive computational trials are conducted on a flat-plate solar air heater and results are compared with those obtained by the other optimization algorithms. For the fair comparison of the TLBO algorithm, the same number of function evaluations are used (Siddhartha, Chauhan, et al., 2012). Population size 30 and maximum number of generations 50 are considered. Like other optimization algorithms (e.g. PSO, ABC, ACO, etc.), TLBO algorithm also does not have any special mechanism to handle the constraints. So, for the constrained optimization problems, it is necessary to incorporate any constraint handling techniques with the TLBO algorithm. In the present experiments, Deb's heuristic constrained handling method (Deb, 2000) is used to handle the constraints with the TLBO algorithm. Deb's method uses a tournament selection operator, in which two solutions are selected and compared with each other. The TLBO code is written in MATLAB and implemented on a laptop having Intel core i3 2.53 GHz processor with 1.85 GB RAM.

Table 1 presents the typical parameter values of solar air heater system (Siddhartha, Sharma, et al., 2012). Table 2 shows the optimum results of thermal performance obtained using the TLBO algorithm and comparison is made with those obtained by the PSO algorithm at $N = 3$ and $S = 600$ W/m^2. The optimum results of thermal performance is also found at different values of N and S, but for the

Table 1. Typical values of solar air heater system parameters (Siddhartha, Chauhan, et al., 2012)

Collector parameters	Values
Length (L) (mm)	1,000
Width (wt) (mm)	200
Height (ht) (mm)	20
Transmittance–absorptance (τa)	0.85
Emissivity of glass cover	0.88
Emissivity of glass plate	0.85–0.95
Tilt angle (β)	$0° \leq \beta \leq 70°$

Table 2. Set of optimal results at N = 3 and S = 600 W/m²

Cases	Algorithms	v	β	ϵ_p	T_a (K)	Temperature rise (K)	ηth (%)
Case 1	PSO (Siddhartha, Sharma, et al., 2012)	1	68.36°	0.89	280.43	10.68	72.42
	TLBO	1.23	59.58°	0.8835	293.93	2.1395	**76.6739**
Case 2	PSO (Siddhartha, Sharma, et al., 2012)	1.02	70.31°	0.94	291.46	10.64	72.19
	TLBO	1.84	69.46°	0.92	294.67	10.62	**76.3181**
Case 3	PSO (Siddhartha, Sharma, et al., 2012)	1.77	70°	0.90	280.01	10.66	72.31
	TLBO	1.98	69.89°	0.91	299.39	10.64	**76.4732**

comparison purpose, it is reported at N = 3 and S = 600 since the results for another settings are not available in Siddhartha, Chauhan, et al. (2012). From Table 2, it can be seen that the thermal efficiency is improved by 5.54% for case 1, 5.39% for case 2, and 5.44% for case 3 using the TLBO algorithm. The optimal thermal performance corresponding to the optimized set of values of velocity (v), tilt angle (β), emissivity of plate (ϵ_p), and ambient temperature (T_a) is determined using the TLBO algorithm as provided in Table 2.

Table 3 presents the range of thermal performance variation for different number of glass cover plates. In total, three sets of glass plates have been considered. Three cases are considered to evaluate the thermal performance of solar air heater using the TLBO algorithm and the results are compared with the PSO algorithm. From Table 3, it can be seen that the thermal efficiency increases as the number of glass cover plate increases. For case 1, the maximum thermal efficiency is obtained at S = 600 and Re = 20,000 and is improved by 8.70, 7.36, and 5.54% for N = 1, 2, and 3, respectively, using the TLBO algorithm. For case 2, the maximum thermal efficiency is obtained at S = 600 and Re = 20,000 and is improved by 8.89, 7.35, and 5.22% for N = 1, 2, and 3, respectively, using the TLBO algorithm. For case 3, the maximum thermal efficiency is obtained at S = 600 and Re = 20,000 and is improved by 8.88, 7.21, and 5.21% for N = 1, 2, and 3, respectively, using TLBO algorithm.

Table 4 presents a set of optimum results of thermal performance of solar air heater at different Reynolds numbers for N = 1 and S = 600 W/m² using the TLBO algorithm and the results are compared with those obtained by GA, PSO, and SA algorithms. The thermal performance in terms of thermal efficiency and different operating parameters for different Reynolds number varying from 2,000 to 20,000 with incremental step of 2,000 are estimated and included in Table 4. In Table 4, the symbol "–" indicates that the results are not available in the cited reference. From Table 4, it can be seen that the thermal efficiency of solar air heater obtained using the TLBO algorithm is better than that obtained using other optimization algorithms by the previous researchers.

Table 3. Range of thermal performance variation for different number of glass cover plates

Cases	Algorithms	N = 1		N = 2		N = 3	
		Min. ηth (%) (S = 1,000, Re = 2,000)	Max. ηth (%) (S = 600, Re = 20,000)	Min. ηth (%) (S = 1,000, Re = 2,000)	Max. ηth (%) (S = 600, Re = 20,000)	Min. ηth (%) (S = 1,000, Re = 2,000)	Max. ηth (%) (S = 600, Re = 20,000)
Case 1	PSO (Siddhartha, Sharma, et al., 2012)	17.24	63.88	22.95	69.08	26.36	72.42
	TLBO	**29.4882**	**69.9757**	**36.1047**	**74.5783**	**41.3158**	**76.6739**
Case 2	PSO (Siddhartha, Sharma, et al., 2012)	17.50	63.15	22.54	68.55	27.32	71.63
	TLBO	**31.1038**	**69.3214**	**35.8726**	**73.9923**	**42.6736**	**75.5839**
Case 3	PSO (Siddhartha, Sharma, et al., 2012)	17.65	62.89	22.91	68.88	27.10	72.31
	TLBO	**31.7482**	**69.0238**	**36.0827**	**74.2482**	**42.8113**	**76.2941**

Table 5 presents the set of optimum results at different Reynolds numbers, $N = 2$ and $S = 600$ W/m². It can be seen that the maximum thermal efficiency of 74.5783% is obtained using the TLBO algorithm at Reynolds number of 20,000, with $v = 1.8635$ m/s, $T_a = 304.6286$ K, $\beta = 6.3283°$, and $\epsilon_p = 0.8738$. The set of optimum results at different Reynolds numbers, $N = 3$ and $S = 600$ W/m² are shown in Table 6. The maximum thermal efficiency obtained using TLBO algorithm is 76.6739%. It can be observed that for the same settings, the thermal performance of a SFPSAH obtained by the TLBO algorithm is better as compared to the other algorithms.

Table 7 presents the set of optimum results at different Reynolds numbers, $N = 1$ and $S = 800$ W/m². It can be seen that the maximum thermal efficiency of 69.8921% is obtained using the TLBO algorithm at Reynolds number of 20,000, with $v = 2.3732$ m/s, $T_a = 305.8392$ K, $\beta = 43.4174°$, and $\epsilon_p = 0.9146$. The set of optimum results at different Reynolds numbers, $N = 2$ and $S = 800$ W/m² is shown in Table 8. The maximum thermal efficiency obtained using TLBO algorithm is 74.4992%. It can be observed that for the same settings, the thermal performance of a SFPSAH obtained by the TLBO algorithm is better as compared to the other algorithms.

A set of optimum results at different Reynolds numbers, $N = 3$ and $S = 800$ W/m² is shown in Table 9. The maximum thermal efficiency obtained using the TLBO algorithm is 76.5913%. It can be observed that for the same settings, the thermal performance of a SFPSAH is better for the TLBO algorithm as compared to that obtained by the other algorithms. Table 10 presents the set of optimum results at different Reynolds numbers, $N = 1$ and $S = 1,000$ W/m². It can be seen that the maximum thermal efficiency of 69.1102% is obtained using the TLBO algorithm at Reynolds number of 20,000 with $v = 2.1194$ m/s, $T_a = 296.9633$ K, $\beta = 36.6728°$, and $\epsilon_p = 0.9043$.

Table 11 presents the set of optimum results at different Reynolds numbers, $N = 2$ and $S = 1,000$ W/m². It can be seen that the maximum thermal efficiency of 74.2392% is obtained using the TLBO algorithm at Reynolds number of 20,000, with $v = 2.3991$ m/s, $T_a = 298.7654$ K, $\beta = 18.3877°$, and $\epsilon_p = 0.9257$. The set of optimum results at different Reynolds numbers, $N = 3$ and $S = 1,000$ W/m² is shown in Table 12. The maximum thermal efficiency obtained using TLBO algorithm is 76.4188%. It can be observed that for the same settings, the thermal performance of a SFPSAH obtained by the TLBO algorithm is better as compared to the other algorithms.

Set of optimum results at different Reynolds numbers, $N = 1$ and $S = 1200$ W/m² is shown in Table 13. The maximum thermal efficiency obtained using TLBO algorithm is 69.06378%. It can be observed that for the same settings, the thermal performance of a SFPSAH is better for the TLBO

S. No.		Re	v	T_a	β	ϵ_p	$T_0 - T_i$	ηth (%)
								Table 4. Set of optimum results at different Reynolds numbers (N = 1 and S = 600 W/m²)
1	GA	2,000	1.0392	301.6078	41.7255	0.8904	7.7582	29.2294
	PSO	2,000	–	–	–	–	–	–
	SA	2,000	–	–	–	–	–	19.5737
	TLBO	2,000	1.3412	293.3784	36.643	0.8826	6.4286	**31.7385**
2	GA	4,000	2.9686	295.1765	57.098	0.8806	5.6814	42.1749
	PSO	4,000	–	–	–	–	–	–
	SA	4,000	–	–	–	–	–	31.9158
	TLBO	4,000	2.3485	297.4782	41.3678	0.8698	5.4386	**43.5603**
3	GA	6,000	1.6745	299.8824	19.2157	0.8751	4.5364	49.7669
	PSO	6,000	–	–	–	–	–	–
	SA	6,000	–	–	–	–	–	40.3917
	TLBO	6,000	1.8692	303.7547	29.3782	0.9173	4.3298	**50.0247**
4	GA	8,000	2.2392	296.3529	25.8039	0.8798	3.7907	55.0673
	PSO	8,000	–	–	–	–	–	–
	SA	8,000	–	–	–	–	–	46.2107
	TLBO	8,000	2.1283	301.8377	45.8828	0.8745	3.7065	**55.9934**
5	GA	10,000	2.7569	302.2353	28	0.8684	3.2359	58.8518
	PSO	10,000	–	–	–	–	–	–
	SA	10,000	–	–	–	–	–	50.9748
	TLBO	10,000	2.2319	299.6739	14.2793	0.8835	3.0531	**60.4672**
6	GA	12,000	1.1412	290.7059	30.7451	0.8578	2.9532	61.7762
	PSO	12,000	–	–	–	–	–	–
	SA	12,000	–	–	–	–	–	54.8929
	TLBO	12,000	1.7835	299.4882	12.4825	0.8621	2.6854	**62.8964**
7	GA	14,000	2.8588	307.3333	42.8235	0.8669	2.5477	63.9401
	PSO	14,000	–	–	–	–	–	–
	SA	14,000	–	–	–	–	–	57.6921
	TLBO	14,000	2.5683	302.4248	48.3732	0.8754	2.4579	**64.7136**
8	GA	16,000	1.2588	302.1569	52.7059	0.8731	2.359	65.9159
	PSO	16,000	–	–	–	–	–	–
	SA	16,000	–	–	–	–	–	60.3319
	TLBO	16,000	1.6735	298.2994	49.3467	0.8634	2.2570	**66.4579**
9	GA	18,000	2.8902	308.3529	61.4902	0.8896	2.1111	67.461
	PSO	18,000	–	–	–	–	–	–
	SA	18,000	–	–	–	–	–	62.3207
	TLBO	18,000	1.9827	304.8321	52.3473	0.8739	2.0966	**68.7547**
10	GA	20,000	1.5725	305.451	39.8039	0.9382	1.9403	68.7416
	PSO	20,000	–	–	–	–	–	63.88
	SA	20,000	–	–	–	–	–	64.0582
	TLBO	20,000	2.1293	302.4858	53.9622	0.8734	1.8965	**69.9757**

Note: The symbol "–" indicates results are not available in the corresponding literature.

Source: The data of the GA, PSO, and SA are taken from Varun and Siddhartha (2010), Siddhartha, Sharma, et al. (2012) and Siddhartha, Chauhan, et al. (2012), respectively.

S. No.		Re	v	T_a	β	ϵ_p	$T_0 - T_i$	ηth (%)

Table 5. Set of optimum results at different Reynolds numbers ($N = 2$ and $S = 600$ W/m²)

S. No.		Re	v	T_a	β	ϵ_p	$T_0 - T_i$	ηth (%)
1	GA	2,000	1.5725	293.3725	54.3529	0.8571	10.2868	36.8908
	PSO	2,000	–	–	–	–	–	–
	SA	2,000	–	–	–	–	–	25.6859
	TLBO	2,000	1.2965	297.9374	37.8745	0.8734	8.9456	**38.1378**
2	GA	4,000	1.3765	301.2157	3.5686	0.9461	6.8671	50.324
	PSO	4,000	–	–	–	–	–	–
	SA	4,000	–	–	–	–	–	39.8566
	TLBO	4,000	1.4385	304.7828	18.6564	0.8935	5.7835	**52.6885**
3	GA	6,000	2.4118	308.4313	31.5686	0.9492	5.239	57.6223
	PSO	6,000	–	–	–	–	–	–
	SA	6,000	–	–	–	–	–	48.6497
	TLBO	6,000	1.8937	305.8827	49.5543	0.9253	4.9831	**58.8843**
4	GA	8,000	1.4	310	56.549	0.8598	4.4204	62.4028
	PSO	8,000	–	–	–	–	–	–
	SA	8,000	–	–	–	–	–	54.6175
	TLBO	8,000	1.5735	309.8372	44.8761	0.9146	3.8736	**63.9765**
5	GA	10,000	2.098	300.5098	3.2941	0.8755	3.7091	65.5798
	PSO	10,000	–	–	–	–	–	–
	SA	10,000	–	–	–	–	–	58.9721
	TLBO	10,000	1.7732	303.4788	21.3567	0.8921	3.5787	**67.5684**
6	GA	12,000	1.8941	300.6667	9.3333	0.8606	3.2273	67.9671
	PSO	12,000	–	–	–	–	–	–
	SA	12,000	–	–	–	–	–	62.1944
	TLBO	12,000	1.3348	302.1784	16.7833	0.8846	3.0174	**68.8355**
7	GA	14,000	2.1843	290.3137	40.3529	0.9206	2.8679	69.6984
	PSO	14,000	–	–	–	–	–	–
	SA	14,000	–	–	–	–	–	64.709
	TLBO	14,000	2.4927	296.3229	34.2849	0.8953	2.6583	**70.3568**
8	GA	16,000	1.7294	298.7843	68.6257	0.8939	2.5842	71.3131
	PSO	16,000	–	–	–	–	–	–
	SA	16,000	–	–	–	–	–	66.7539
	TLBO	16,000	1.8943	301.1673	45.8392	0.8635	2.4846	**72.3462**
9	GA	18,000	1.0471	302	23.0588	0.8888	2.318	72.4333
	PSO	18,000	–	–	–	–	–	–
	SA	18,000	–	–	–	–	–	68.3109
	TLBO	18,000	1.7139	303.7845	11.2863	0.9265	2.2739	**73.1954**
10	GA	20,000	2.1608	299.8039	11.5294	0.8939	2.1088	73.4772
	PSO	20,000	–	–	–	–	–	69.08
	SA	20,000	–	–	–	–	–	69.7965
	TLBO	20,000	1.8635	304.6286	6.3283	0.8738	2.0683	**74.5783**

Note: The symbol "–" indicates results are not available in the corresponding literature.

Source: The data of the GA, PSO, and SA are taken from Varun and Siddhartha (2010), Siddhartha, Sharma, et al. (2012) and Siddhartha, Chauhan, et al. (2012), respectively.

Table 6. Set of optimum results at different Reynolds numbers (N = 3 and S = 600 W/m²)

S. No.		Re	v	T_a	β	ϵ_p	$T_0 - T_i$	ηth (%)
1	GA	2,000	1.4	309.2941	41.1765	0.8708	11.35	41.7897
	PSO	2,000	1	282.76	14.4	0.92	41.69	28.38
	SA	2,000	–	–	–	–	–	19.5737
	TLBO	2,000	1.1265	301.3783	22.6394	0.8856	9.2992	**43.3532**
2	GA	4,000	2.2627	293.3725	48.8627	0.8908	7.8267	55.1325
	PSO	4,000	1	282.38	23.1	0.86	31.89	43.37
	SA	4,000	–	–	–	–	–	31.9158
	TLBO	4,000	1.4826	299.5738	36.8323	0.8912	7.2369	**56.7543**
3	GA	6,000	1.1961	292.7451	64.5098	0.9488	5.9713	61.852
	PSO	6,000	1	280.04	53.78	0.9	25.95	52.88
	SA	6,000	–	–	–	–	–	40.3917
	TLBO	6,000	1.6357	294.4782	45.2376	0.9243	4.9832	**63.4863**
4	GA	8,000	1.9725	292.2745	0	0.8633	4.7422	66.1788
	PSO	8,000	1.02	280.07	19.46	0.9	21.28	58.25
	SA	8,000	–	–	–	–	–	46.2107
	TLBO	8,000	1.3789	290.7489	31.5943	0.9074	4.1294	**67.6733**
5	GA	10,000	2.7098	295.5686	2.7451	0.8916	3.9458	68.8946
	PSO	10,000	1	280.11	30.79	0.93	18.23	61.95
	SA	10,000	–	–	–	–	–	50.9748
	TLBO	10,000	1.4937	291.7112	42.6582	0.9247	3.7836	**70.4711**
6	GA	12,000	1.1961	298.2353	49.9608	0.8516	3.4344	71.0254
	PSO	12,000	1	280.04	62.86	0.94	16.08	65.49
	SA	12,000	–	–	–	–	–	54.8929
	TLBO	12,000	1.7619	300.1388	57.9334	0.9376	3.2694	**72.2885**
7	GA	14,000	2.0275	292.2745	66.7059	0.9088	3.0221	72.5723
	PSO	14,000	1	280.02	20.45	0.93	14.13	67.72
	SA	14,000	–	–	–	–	–	57.6921
	TLBO	14,000	1.2834	295.7835	39.1178	0.9421	2.8479	**73.6319**
8	GA	16,000	2.3725	301.2157	65.6078	0.9461	2.6706	73.8123
	PSO	16,000	1.04	280.01	44.17	0.9	12.76	69.28
	SA	16,000	–	–	–	–	–	60.3319
	TLBO	16,000	1.7482	298.5294	61.9326	0.8995	2.5636	**74.3058**
9	GA	18,000	1.4038	297.6078	20.8627	0.879	2.4098	74.8367
	PSO	18,000	1	280.02	26.52	0.87	11.57	70.84
	SA	18,000	–	–	–	–	–	62.3207
	TLBO	18,000	1.1619	293.8943	21.8311	0.9105	2.3295	**75.9472**
10	GA	20,000	2.9529	296.1176	65.3333	0.8618	2.2047	75.6454
	PSO	20,000	1	280.43	68.36	0.89	10.68	72.42
	SA	20,000	–	–	–	–	–	64.0582
	TLBO	20,000	1.2729	293.9362	59.5832	0.8835	2.1395	**76.6739**

Note: The symbol "–" indicates results are not available in the corresponding literature.

Source: The data of the GA, PSO, and SA are taken from Varun and Siddhartha (2010), Siddhartha, Sharma, et al. (2012) and Siddhartha, Chauhan, et al. (2012), respectively.

S. No.		Re	v	T_a	β	ϵ_p	$T_0 - T_i$	ηth (%)
Table 7. Set of optimum results at different Reynolds numbers ($N = 1$ and $S = 800$ W/m²)								
1	GA	2,000	2.7725	292.0392	24.4314	0.9229	9.2733	28.2512
	PSO	2,000	–	–	–	–	–	–
	SA	2,000	–	–	–	–	–	18.7072
	TLBO	2,000	2.1139	297.3844	38.3772	0.8895	7.8746	**30.9532**
2	GA	4,000	1.2588	296.0392	53.5294	0.9237	7.5112	41.1115
	PSO	4,000	–	–	–	–	–	–
	SA	4,000	–	–	–	–	–	30.836
	TLBO	4,000	1.6726	294.5783	23.4788	0.8631	6.5937	**43.0947**
3	GA	6,000	2.098	294.2353	6.8627	0.9135	5.8441	48.9045
	PSO	6,000	–	–	–	–	–	–
	SA	6,000	–	–	–	–	–	39.2944
	TLBO	6,000	1.8371	290.1247	23.9984	0.8953	5.3693	**49.9363**
4	GA	8,000	2.7176	290.7059	59.0196	0.9331	4.9948	54.1312
	PSO	8,000	–	–	–	–	–	–
	SA	8,000	–	–	–	–	–	45.5684
	TLBO	8,000	2.2193	298.3667	47.2787	0.9256	4.6932	**55.3843**
5	GA	10,000	1.7294	301.451	57.098	0.881	4.3663	58.2862
	PSO	10,000	–	–	–	–	–	–
	SA	10,000	–	–	–	–	–	50.2908
	TLBO	10,000	1.5632	299.4882	34.5367	0.9173	4.1395	**59.6832**
6	GA	12,000	1.9176	300.3529	48.0392	0.9069	3.8012	61.0959
	PSO	12,000	–	–	–	–	–	–
	SA	12,000	–	–	–	–	–	54.1293
	TLBO	12,000	1.8936	302.4781	38.4673	0.9268	3.6395	**62.8734**
7	GA	14,000	1.7765	297.451	57.6471	0.8947	3.4416	63.5274
	PSO	14,000	–	–	–	–	–	–
	SA	14,000	–	–	–	–	–	57.1511
	TLBO	14,000	1.4423	299.4268	27.5764	0.8748	3.1957	**64.2154**
8	GA	16,000	3.0	291.6471	66.1569	0.9473	3.0741	65.3202
	PSO	16,000	–	–	–	–	–	–
	SA	16,000	–	–	–	–	–	59.6272
	TLBO	16,000	2.7394	295.3278	51.6478	0.9376	2.8643	**66.0115**
9	GA	18,000	2.5922	291.4118	67.7647	0.9214	2.8268	66.943
	PSO	18,000	–	–	–	–	–	–
	SA	18,000	–	–	–	–	–	61.4528
	TLBO	18,000	2.8846	298.2783	32.9754	0.9475	2.6493	**68.0364**
10	GA	20,000	2.0353	309.2941	14.2745	0.8606	2.5614	68.2924
	PSO	20,000	–	–	–	–	–	–
	SA	20,000	–	–	–	–	–	63.3172
	TLBO	20,000	2.3732	305.8392	43.4174	0.9146	2.4395	**69.8921**

Note: The symbol "–" indicates results are not available in the corresponding literature.

Source: The data of the GA, PSO, and SA are taken from Varun and Siddhartha (2010), Siddhartha, Sharma, et al. (2012) and Siddhartha, Chauhan, et al. (2012), respectively.

S. No.		Re	v	T_a	β	ϵ_p	$T_0 - T_i$	ηth (%)
1	GA	2,000	1.1098	308.7451	26.0784	0.9186	12.4267	35.6044
	PSO	2,000	–	–	–	–	–	–
	SA	2,000	–	–	–	–	–	24.5662
	TLBO	2,000	1.3847	301.2184	35.7628	0.8845	11.5783	**37.1049**
2	GA	4,000	2.2392	294.7843	34.3137	0.9229	9.0186	49.2273
	PSO	4,000	–	–	–	–	–	–
	SA	4,000	–	–	–	–	–	38.3891
	TLBO	4,000	2.8374	291.9345	52.8593	0.8936	8.3491	**50.6738**
3	GA	6,000	2.8039	297.7647	26.6275	0.9229	6.9579	56.7569
	PSO	6,000	–	–	–	–	–	–
	SA	6,000	–	–	–	–	–	47.5704
	TLBO	6,000	2.3948	301.8943	41.5675	0.8954	6.1103	**58.2194**
4	GA	8,000	2.5529	292.1176	1.098	0.8673	5.7996	61.6262
	PSO	8,000	–	–	–	–	–	–
	SA	8,000	–	–	–	–	–	53.7162
	TLBO	8,000	2.1284	296.3848	18.3573	0.8843	5.2295	**63.3042**
5	GA	10,000	2.6	290.4706	17.5686	0.8884	4.9222	65.0241
	PSO	10,000	–	–	–	–	–	–
	SA	10,000	–	–	–	–	–	58.1612
	TLBO	10,000	2.1839	297.8835	5.6884	0.9256	4.6402	**66.8211**
6	GA	12,000	1.5255	303.4902	48.3137	0.852	4.2986	67.4832
	PSO	12,000	–	–	–	–	–	–
	SA	12,000	–	–	–	–	–	61.3731
	TLBO	12,000	1.3111	298.7145	12.4678	0.8853	3.9836	**68.7343**
7	GA	14,000	1.1412	290.0784	11.2549	0.9488	3.7964	69.3503
	PSO	14,000	–	–	–	–	–	–
	SA	14,000	–	–	–	–	–	64.1098
	TLBO	14,000	1.7829	294.7883	21.4573	0.9145	3.5739	**70.2859**
8	GA	16,000	1.0314	294.7059	25.8039	0.8869	3.4103	70.8033
	PSO	16,000	–	–	–	–	–	–
	SA	16,000	–	–	–	–	–	66.1263
	TLBO	16,000	1.4388	290.8253	34.8754	0.9354	3.2954	**71.9348**
9	GA	18,000	1.7686	309.451	5.4902	0.8912	3.0433	72.1471
	PSO	18,000	–	–	–	–	–	–
	SA	18,000	–	–	–	–	–	67.8595
	TLBO	18,000	1.9999	302.7843	21.5784	0.9257	2.9184	**73.7721**
10	GA	20,000	1.0941	301.2157	21.1373	0.9382	2.8031	73.1107
	PSO	20,000	–	–	–	–	–	–
	SA	20,000	–	–	–	–	–	69.3062
	TLBO	20,000	1.6583	307.2738	7.8345	0.8853	2.6754	**74.4992**

Table 8. Set of optimum results at different Reynolds numbers (N = 2 and S = 800 W/m²)

Note: The symbol "–" indicates results are not available in the corresponding literature.

Source: The data of the GA, PSO, and SA are taken from Varun and Siddhartha (2010), Siddhartha, Sharma, et al. (2012) and Siddhartha, Chauhan, et al. (2012), respectively.

S. No.		Re	v	T_a	β	ϵ_p	$T_0 - T_i$	ηth (%)
Table 9. Set of optimum results at different Reynolds numbers (N = 3 and S = 800 W/m²)								
1	GA	2,000	2.3804	299.3333	12.902	0.9229	14.4294	40.3751
	PSO	2,000	–	–	–	–	–	–
	SA	2,000	–	–	–	–	–	18.7072
	TLBO	2,000	1.8947	293.8122	32.3632	0.8954	12.4839	**41.8493**
2	GA	4,000	1.8078	298.7059	49.6863	0.9088	10.1644	54.124
	PSO	4,000	–	–	–	–	–	–
	SA	4,000	–	–	–	–	–	30.836
	TLBO	4,000	2.1378	293.7229	21.4673	0.8842	8.9937	**55.9932**
3	GA	6,000	2.7725	299.098	29.6471	0.8524	7.694	61.1521
	PSO	6,000	–	–	–	–	–	–
	SA	6,000	–	–	–	–	–	39.2944
	TLBO	6,000	2.4296	290.1283	36.9761	0.8824	6.8746	**62.4839**
4	GA	8,000	2.3412	307.098	70	0.939	6.2278	65.5463
	PSO	8,000	–	–	–	–	–	–
	SA	8,000	–	–	–	–	–	45.5684
	TLBO	8,000	2.6510	304.9392	58.3468	0.8964	5.7385	**67.2395**
5	GA	10,000	1.4863	293.5294	14.8235	0.8751	5.2581	68.485
	PSO	10,000	–	–	–	–	–	–
	SA	10,000	–	–	–	–	–	50.2908
	TLBO	10,000	1.5738	301.3935	34.4674	0.9365	4.8624	**69.8883**
6	GA	12,000	1.6039	296.4314	45.2941	0.941	4.5261	70.5836
	PSO	12,000	–	–	–	–	–	–
	SA	12,000	–	–	–	–	–	54.1293
	TLBO	12,000	1.2399	300.3782	25.2485	0.9156	4.3638	**72.0012**
7	GA	14,000	1.7059	309.6863	21.9608	0.8614	3.9446	72.2261
	PSO	14,000	–	–	–	–	–	–
	SA	14,000	–	–	–	–	–	57.1511
	TLBO	14,000	1.3911	304.5692	9.2474	0.8951	3.7953	**73.4924**
8	GA	16,000	1.4627	305.7647	61.2157	0.9124	3.5459	73.5504
	PSO	16,000	–	–	–	–	–	–
	SA	16,000	–	–	–	–	–	59.6272
	TLBO	16,000	1.6784	299.7832	48.3466	0.8627	3.3681	**74.1775**
9	GA	18,000	2.4824	296.2745	44.1961	0.9331	3.1948	74.5676
	PSO	18,000	–	–	–	–	–	–
	SA	18,000	–	–	–	–	–	61.4528
	TLBO	18,000	2.0384	291.5622	29.4673	0.8776	3.0193	**75.3221**
10	GA	20,000	1.2275	309.2157	51.8824	0.9422	2.9059	75.3931
	PSO	20,000	–	–	–	–	–	–
	SA	20,000	–	–	–	–	–	63.3172
	TLBO	20,000	1.6293	302.7223	32.4672	0.9189	2.7646	**76.5913**

Note: The symbol "–" indicates results are not available in the corresponding literature.

Source: The data of the GA, PSO, and SA are taken from Varun and Siddhartha (2010), Siddhartha, Sharma, et al. (2012) and Siddhartha, Chauhan, et al. (2012), respectively.

S. No.		Re	v	T_a	β	ϵ_p	$T_0 - T_i$	ηth (%)
								Table 10. Set of optimum results at different Reynolds numbers (N = 1 and S = 1,000 W/m^2)
1	GA	2,000	2.2471	298.9412	35.6863	0.8665	11.444	27.4864
	PSO	2,000	–	–	–	–	–	17.24
	SA	2,000	–	–	–	–	–	17.6966
	TLBO	2,000	1.8831	293.2967	47.3782	0.8965	10.3842	**29.4882**
2	GA	4,000	2.3176	291.4118	10.4314	0.8539	8.9379	40.1084
	PSO	4,000	–	–	–	–	–	–
	SA	4,000	–	–	–	–	–	30.0172
	TLBO	4,000	2.1293	295.2434	4.8932	0.8834	7.8343	**41.9837**
3	GA	6,000	2.6941	298.3137	4.3922	0.85	7.1593	48.1196
	PSO	6,000	–	–	–	–	–	–
	SA	6,000	–	–	–	–	–	38.1824
	TLBO	6,000	2.1934	299.8387	19.3893	0.8924	6.9837	**49.8212**
4	GA	8,000	1.2667	303.4902	1.9216	0.932	6.0706	53.5817
	PSO	8,000	–	–	–	–	–	–
	SA	8,000	–	–	–	–	–	44.6621
	TLBO	8,000	1.3847	297.3289	18.4923	0.8999	5.7184	**55.2149**
5	GA	10,000	1.8471	295.2549	26.3529	0.8633	5.3605	57.657
	PSO	10,000	–	–	–	–	–	–
	SA	10,000	–	–	–	–	–	49.3149
	TLBO	10,000	2.2389	299.5263	43.6722	0.9135	4.9285	**59.1038**
6	GA	12,000	1.3765	304.7451	22.2353	0.9484	4.6377	60.5783
	PSO	12,000	–	–	–	–	–	–
	SA	12,000	–	–	–	–	–	53.4182
	TLBO	12,000	1.2184	300.3784	11.3832	0.9145	4.3795	**62.4895**
7	GA	14,000	2.7725	309.2941	17.8431	0.9288	4.0572	63.0055
	PSO	14,000	–	–	–	–	–	–
	SA	14,000	–	–	–	–	–	56.5065
	TLBO	14,000	2.3495	303.5638	36.2781	0.8936	3.8947	**64.1783**
8	GA	16,000	2.1451	293.3725	5.2157	0.9371	3.7709	65.0191
	PSO	16,000	–	–	–	–	–	–
	SA	16,000	–	–	–	–	–	59.0541
	TLBO	16,000	2.4183	297.4567	23.6638	0.8954	3.6397	**65.9987**
9	GA	18,000	1.7529	291.2549	27.1765	0.9127	3.4971	66.6081
	PSO	18,000	–	–	–	–	–	–
	SA	18,000	–	–	–	–	–	61.1254
	TLBO	18,000	2.0847	290.8743	8.2563	0.8845	3.3865	**67.6341**
10	GA	20,000	2.4039	292.1961	55.7255	0.9182	3.2241	67.7749
	PSO	20,000	–	–	–	–	–	–
	SA	20,000	–	–	–	–	–	62.9629
	TLBO	20,000	2.1194	296.9633	36.6728	0.9043	3.1975	**69.1102**

Note: The symbol "–" indicates results are not available in the corresponding literature.

Source: The data of the GA, PSO, and SA are taken from Varun and Siddhartha (2010), Siddhartha, Sharma, et al. (2012) and Siddhartha, Chauhan, et al. (2012), respectively.

Table 11. Set of optimum results at different Reynolds numbers ($N = 2$ and $S = 1{,}000$ W/m²)							
S. No.		Re	v	T_a	β	ϵ_p	$T_0 - T_i$	ηth (%)
1	GA	2,000	2.9294	308.8235	27.1765	0.9249	14.3719	34.5239
	PSO	2,000	–	–	–	–	–	22.95
	SA	2,000	–	–	–	–	–	23.5642
	TLBO	2,000	2.7839	302.5673	39.2781	0.9054	13.7954	**36.1047**
2	GA	4,000	2.0196	296.5098	11.5294	0.8696	11.0482	48.311
	PSO	4,000	–	–	–	–	–	–
	SA	4,000	–	–	–	–	–	37.5379
	TLBO	4,000	2.7329	301.3485	29.4882	0.8756	9.9535	**49.8348**
3	GA	6,000	2.6706	295.8824	67.2549	0.941	8.7547	56.0124
	PSO	6,000	–	–	–	–	–	–
	SA	6,000	–	–	–	–	–	46.5347
	TLBO	6,000	2.2937	300.3962	48.3257	0.8842	8.1482	**57.9234**
4	GA	8,000	1.5333	292.3529	9.6078	0.9245	7.1766	61.0463
	PSO	8,000	–	–	–	–	–	–
	SA	8,000	–	–	–	–	–	52.6971
	TLBO	8,000	1.6621	290.4852	27.6829	0.8944	6.8637	**62.8342**
5	GA	10,000	2.2627	303.4118	52.7059	0.8755	6.0997	64.3693
	PSO	10,000	–	–	–	–	–	–
	SA	10,000	–	–	–	–	–	57.3815
	TLBO	10,000	2.0382	297.3584	32.5710	0.8685	5.6786	**66.1038**
6	GA	12,000	2.5765	303.1961	67.2549	0.9429	5.2672	67.0317
	PSO	12,000	–	–	–	–	–	–
	SA	12,000	–	–	–	–	–	60.7765
	TLBO	12,000	2.2146	302.2468	53.2892	0.8821	4.8953	**68.8937**
7	GA	14,000	1.4549	303.6471	5.2157	0.9029	4.6683	69.0025
	PSO	14,000	–	–	–	–	–	–
	SA	14,000	–	–	–	–	–	63.3947
	TLBO	14,000	1.7816	294.6549	32.8267	0.9421	4.4168	**70.0127**
8	GA	16,000	2.1686	307.8824	56	0.8692	4.2096	70.5238
	PSO	16,000	–	–	–	–	–	–
	SA	16,000	–	–	–	–	–	65.6112
	TLBO	16,000	1.9725	302.3473	34.9392	0.9184	3.9994	**71.6739**
9	GA	18,000	1.7294	304.1961	65.0588	0.9375	3.8271	71.7789
	PSO	18,000	–	–	–	–	–	–
	SA	18,000	–	–	–	–	–	67.4154
	TLBO	18,000	1.6427	308.3697	47.1003	0.8756	3.6379	**72.5611**
10	GA	20,000	2.0196	302.4706	4.6667	0.9492	3.4616	72.8854
	PSO	20,000	–	–	–	–	–	–
	SA	20,000	–	–	–	–	–	68.8388
	TLBO	20,000	2.3991	298.7654	18.3877	0.9257	3.3982	**74.2392**

Note: The symbol "–" indicates results are not available in the corresponding literature.

Source: The data of the GA, PSO, and SA are taken from Varun and Siddhartha (2010), Siddhartha, Sharma, et al. (2012) and Siddhartha, Chauhan, et al. (2012), respectively.

S. No.		Re	v	T_a	β	ϵ_p	$T_0 - T_i$	ηth (%)

Table 12. Set of optimum results at different Reynolds numbers (N = 3 and S = 1,000 W/m²)

S. No.		Re	v	T_a	β	ϵ_p	$T_0 - T_i$	ηth (%)
1	GA	2,000	2.2314	296.8235	69.451	0.9073	18.1867	39.2811
	PSO	2,000	–	–	–	–	–	26.36
	SA	2,000	–	–	–	–	–	17.6966
	TLBO	2,000	1.8366	299.3458	56.2886	0.9256	16.7855	**41.3158**
2	GA	4,000	1.5725	303.098	47.7647	0.8614	12.4564	53.1766
	PSO	4,000	–	–	–	–	–	–
	SA	4,000	–	–	–	–	–	30.0172
	TLBO	4,000	1.3857	300.4832	34.5673	0.8953	11.2398	**55.0012**
3	GA	6,000	2.1451	304.5882	29.098	0.888	9.4275	60.5002
	PSO	6,000	–	–	–	–	–	–
	SA	6,000	–	–	–	–	–	38.1824
	TLBO	6,000	1.8453	307.9765	16.4873	0.8745	8.6536	**62.2119**
4	GA	8,000	1.9333	301.1373	48.3137	0.8739	7.7438	64.9469
	PSO	8,000	–	–	–	–	–	–
	SA	8,000	–	–	–	–	–	44.6621
	TLBO	8,000	2.1183	296.3696	62.5893	0.8951	7.0964	**66.3957**
5	GA	10,000	2.4431	294.3922	24.9804	0.9175	6.4845	68.0499
	PSO	10,000	–	–	–	–	–	–
	SA	10,000	–	–	–	–	–	49.3149
	TLBO	10,000	2.1827	298.4594	8.3462	0.9054	5.996	**69.7367**
6	GA	12,000	2.1294	301.7647	33.4902	0.8904	5.5936	70.27
	PSO	12,000	–	–	–	–	–	–
	SA	12,000	–	–	–	–	–	53.4182
	TLBO	12,000	2.6748	297.2470	48.3672	0.8797	5.3478	**71.7732**
7	GA	14,000	1.0627	303.4118	3.0196	0.9014	4.9224	71.9223
	PSO	14,000	–	–	–	–	–	–
	SA	14,000	–	–	–	–	–	56.5065
	TLBO	14,000	1.3882	297.6738	23.6482	0.9262	4.6855	**73.3937**
8	GA	16,000	1.8	295.2549	7.1373	0.8782	4.4093	73.2418
	PSO	16,000	–	–	–	–	–	–
	SA	16,000	–	–	–	–	–	59.0541
	TLBO	16,000	1.5638	299.4537	19.3678	0.8963	4.2378	**74.1189**
9	GA	18,000	1.4941	290.7059	58.1961	0.8594	4.0209	74.2811
	PSO	18,000	–	–	–	–	–	–
	SA	18,000	–	–	–	–	–	61.1254
	TLBO	18,000	1.2811	293.2882	23.7748	0.8848	3.8267	**75.1932**
10	GA	20,000	1	291.3333	21.6863	0.9433	3.6408	75.2149
	PSO	20,000	–	–	–	–	–	–
	SA	20,000	–	–	–	–	–	62.3454
	TLBO	20,000	1.3294	294.9836	47.2782	0.9145	3.5796	**76.4188**

Note: The symbol "–" indicates results are not available in the corresponding literature.

Source: The data of the GA, PSO, and SA are taken from Varun and Siddhartha (2010), Siddhartha, Sharma, et al. (2012) and Siddhartha, Chauhan, et al. (2012), respectively.

S. No.		Re	v	T_a	β	ϵ_p	$T_0 - T_i$	ηth (%)
Table 13. Set of optimum results at different Reynolds numbers ($N = 1$ and $S = 1,200$ W/m²)								
1	GA	2,000	1.1098	298	12.0784	0.8524	13.7712	26.6961
	PSO	2,000	–	–	–	–	–	–
	SA	2,000	–	–	–	–	–	17.4604
	TLBO	2,000	1.5248	294.7832	23.3772	0.8738	12.7467	**28.3282**
2	GA	4,000	2.6784	301.8431	30.7451	0.8669	10.178	39.3842
	PSO	4,000	–	–	–	–	–	–
	SA	4,000	–	–	–	–	–	29.0595
	TLBO	4,000	2.4261	296.3782	49.2628	0.8634	9.4684	**41.6739**
3	GA	6,000	2.0431	304.4314	68.3529	0.9022	8.6299	47.2705
	PSO	6,000	–	–	–	–	–	–
	SA	6,000	–	–	–	–	–	37.6487
	TLBO	6,000	1.8131	301.7223	57.2837	0.8854	7.9854	**49.7638**
4	GA	8,000	1.7765	294.7843	58.7451	0.8947	7.4196	52.8345
	PSO	8,000	–	–	–	–	–	–
	SA	8,000	–	–	–	–	–	43.9187
	TLBO	8,000	1.3989	297.3629	62.4629	0.9252	6.8594	**54.7834**
5	GA	10,000	1.4941	297.2941	64.7843	0.8947	6.4737	57.0424
	PSO	10,000	–	–	–	–	–	–
	SA	10,000	–	–	–	–	–	48.8642
	TLBO	10,000	1.7712	301.8264	48.2563	0.9065	5.8832	**59.1021**
6	GA	12,000	1.9333	309.6863	53.2549	0.8653	5.5802	60.1529
	PSO	12,000	–	–	–	–	–	–
	SA	12,000	–	–	–	–	–	52.7259
	TLBO	12,000	2.2316	302.3782	35.2645	0.8812	5.2394	**61.8847**
7	GA	14,000	2.3333	300.2745	23.6078	0.8739	4.9692	62.4509
	PSO	14,000	–	–	–	–	–	–
	SA	14,000	–	–	–	–	–	55.7717
	TLBO	14,000	2.4712	298.3782	43.2842	0.8848	4.7183	**63.9456**
8	GA	16,000	1.3608	302.4706	21.9608	0.8598	4.5627	64.4455
	PSO	16,000	–	–	–	–	–	–
	SA	16,000	–	–	–	–	–	58.3572
	TLBO	16,000	1.7629	299.2774	36.6721	0.8932	4.3752	**65.7832**
9	GA	18,000	2.7255	300.5098	24.1569	0.8951	4.0956	66.2157
	PSO	18,000	–	–	–	–	–	–
	SA	18,000	–	–	–	–	–	60.3263
	TLBO	18,000	2.0527	303.2567	13.8288	0.9382	3.9611	**67.4527**
10	GA	20,000	1.1961	301.451	55.1765	0.9006	3.8724	67.5588
	PSO	20,000	–	–	–	–	–	–
	SA	20,000	–	–	–	–	–	62.3454
	TLBO	20,000	1.2836	304.8453	29.2752	0.9161	3.7923	**69.06378**

Note: The symbol "–" indicates results are not available in the corresponding literature.

Source: The data of the GA, PSO, and SA are taken from Varun and Siddhartha (2010), Siddhartha, Sharma, et al. (2012) and Siddhartha, Chauhan, et al. (2012), respectively.

S. No.		Re	v	T_a	β	ϵ_p	$T_0 - T_i$	ηth (%)
Table 14. Set of optimum results at different Reynolds numbers (N = 2 and S = 1,200 W/m²)								
1	GA	2,000	1.0471	301.7647	33.4902	0.8873	17.796	33.6309
	PSO	2,000	–	–	–	–	–	–
	SA	2,000	–	–	–	–	–	22.6143
	TLBO	2,000	1.3772	298.5773	21.6737	0.8934	15.3859	**35.4562**
2	GA	4,000	2.1451	299.2549	24.9804	0.9339	12.7697	47.4312
	PSO	4,000	–	–	–	–	–	–
	SA	4,000	–	–	–	–	–	36.5246
	TLBO	4,000	2.4436	295.7382	42.8322	0.9132	11.6380	**49.4673**
3	GA	6,000	1.8392	293.8431	40.0784	0.8947	10.3696	55.2555
	PSO	6,000	–	–	–	–	–	–
	SA	6,000	–	–	–	–	–	45.6225
	TLBO	6,000	1.5263	290.3843	65.3738	0.9028	9.3496	**56.9932**
4	GA	8,000	2.0431	306.2353	61.2157	0.9233	8.4761	60.3607
	PSO	8,000	–	–	–	–	–	–
	SA	8,000	–	–	–	–	–	51.9252
	TLBO	8,000	1.8127	302.6325	52.4771	0.8992	7.8913	**62.6748**
5	GA	10,000	1.6431	302.3137	4.6667	0.9139	7.1846	64.0505
	PSO	10,000	–	–	–	–	–	–
	SA	10,000	–	–	–	–	–	56.4899
	TLBO	10,000	1.2328	297.1183	21.3392	0.8846	6.9964	**65.8992**
6	GA	12,000	1.0863	303.9608	56.549	0.8567	6.3845	66.5759
	PSO	12,000	–	–	–	–	–	–
	SA	12,000	–	–	–	–	–	60.0281
	TLBO	12,000	1.3823	299.7837	23.5732	0.8628	5.8732	**68.2184**
7	GA	14,000	2.8745	296.9804	59.8431	0.8547	5.6448	68.6168
	PSO	14,000	–	–	–	–	–	–
	SA	14,000	–	–	–	–	–	62.9441
	TLBO	14,000	2.6122	299.3822	47.2685	0.9163	5.3181	**69.9247**
8	GA	16,000	1.6118	295.9608	34.0392	0.879	5.0491	70.2132
	PSO	16,000	–	–	–	–	–	–
	SA	16,000	–	–	–	–	–	65.1525
	TLBO	16,000	1.2178	302.5732	21.4778	0.8927	4.8832	**71.4692**
9	GA	18,000	1.2353	301.6078	23.6078	0.8684	4.5681	71.4894
	PSO	18,000	–	–	–	–	–	–
	SA	18,000	–	–	–	–	–	67.0587
	TLBO	18,000	1.5721	304.9956	8.3772	0.9027	4.4616	**72.3785**
10	GA	20,000	1.5961	292.8235	21.9608	0.8547	4.1981	72.5851
	PSO	20,000	–	–	–	–	–	–
	SA	20,000	–	–	–	–	–	68.5195
	TLBO	20,000	1.8973	290.3753	16.4672	0.8623	4.1109	**74.1743**

Note: The symbol "–" indicates results are not available in the corresponding literature.

Source: The data of the GA, PSO, and SA are taken from Varun and Siddhartha (2010), Siddhartha, Sharma, et al. (2012) and Siddhartha, Chauhan, et al. (2012), respectively.

S. No.		Re	v	T_a	β	ϵ_p	$T_0 - T_i$	ηth (%)
Table 15. Set of optimum results at different Reynolds numbers (N = 3 and S = 1,200 W/m²)								
1	GA	2,000	1.2431	297.0588	61.4902	0.8602	21.3196	38.1579
	PSO	2,000	–	–	–	–	–	–
	SA	2,000	–	–	–	–	–	17.4604
	TLBO	2,000	1.0151	294.3668	45.3568	0.8926	17.3942	**40.2748**
2	GA	4,000	1.8784	301.6078	12.6275	0.8645	14.505	52.4193
	PSO	4,000	–	–	–	–	–	–
	SA	4,000	–	–	–	–	–	29.0595
	TLBO	4,000	1.6714	297.6782	27.3564	0.8849	12.8242	**54.2748**
3	GA	6,000	1.3216	294.4706	64.2353	0.8594	11.5417	59.8554
	PSO	6,000	–	–	–	–	–	–
	SA	6,000	–	–	–	–	–	37.6487
	TLBO	6,000	1.7893	291.5683	56.7544	0.8941	10.5689	**61.4773**
4	GA	8,000	2.7725	294.6275	32.6667	0.9473	9.1277	64.5362
	PSO	8,000	–	–	–	–	–	–
	SA	8,000	–	–	–	–	–	43.9187
	TLBO	8,000	2.4844	298.4632	12.7348	0.9147	8.7635	**66.2726**
5	GA	10,000	2.5608	302.4706	58.1961	0.8825	7.7548	67.592
	PSO	10,000	–	–	–	–	–	–
	SA	10,000	–	–	–	–	–	48.8642
	TLBO	10,000	2.8637	298.3570	63.4563	0.9123	7.1074	**69.3674**
6	GA	12,000	2.9686	290.7059	43.9216	0.8908	6.7211	69.9637
	PSO	12,000	–	–	–	–	–	–
	SA	12,000	–	–	–	–	–	52.7259
	TLBO	12,000	2.8868	293.9874	56.7432	0.9038	6.3987	**71.2371**
7	GA	14,000	1.2118	293.7647	57.098	0.8986	5.9539	71.6221
	PSO	14,000	–	–	–	–	–	–
	SA	14,000	–	–	–	–	–	55.7717
	TLBO	14,000	1.6585	291.3683	34.8643	0.9142	5.5853	**72.6738**
8	GA	16,000	2.6314	309.2157	16.4706	0.8535	5.225	72.9702
	PSO	16,000	–	–	–	–	–	–
	SA	16,000	–	–	–	–	–	58.3572
	TLBO	16,000	2.8726	305.3783	31.4786	0.8736	4.9643	**73.9738**
9	GA	18,000	2.7647	299.098	44.1961	0.9025	4.7531	74.0904
	PSO	18,000	–	–	–	–	–	–
	SA	18,000	–	–	–	–	–	60.3263
	TLBO	18,000	2.2169	303.4577	29.8642	0.8856	4.5775	**74.8992**
10	GA	20,000	2.8353	306.7843	7.9608	0.9406	4.2964	74.969
	PSO	20,000	–	–	–	–	–	–
	SA	20,000	–	–	–	–	–	62.3454
	TLBO	20,000	2.7825	301.9751	24.5735	0.8821	4.2063	**76.0374**

Note: The symbol "–" indicates results are not available in the corresponding literature.

Source: The data of the GA, PSO, and SA are taken from Varun and Siddhartha (2010), Siddhartha, Sharma, et al. (2012) and Siddhartha, Chauhan, et al. (2012), respectively.

algorithm as compared to other algorithms. Table 14 presents the set of optimum results at different Reynolds numbers, $N = 2$ and $S = 1200$ W/m^2. It can be seen that the maximum thermal efficiency of 74.1743% is obtained using the TLBO algorithm at Reynolds number 20,000, with $v = 1.8973$ m/s, $T_a = 290.3753$ K, $\beta = 16.4672°$, and $\epsilon_p = 0.8623$. A set of optimum results at different Reynolds numbers, $N = 3$ and $S = 1200$ W/m^2 is shown in Table 15. The maximum thermal efficiency obtained using TLBO algorithm is 76.0374%. It can be observed that for the same settings, the thermal performance of a SFPSAH obtained by the TLBO algorithm is better as compared to the other algorithms.

In Tables 5–15, the symbol "–" indicates that the results are not available in the cited reference. From Tables 4–15, it can be seen that TLBO algorithm performed better than the other optimization algorithms considered by the previous researchers.

4.1. Effect of Reynolds number on thermal performance

In this section, the effect of Reynolds number on thermal performance of SFPSAH is analyzed with respect to the design and operating parameters. Figure 3 presents variation of thermal performance for different Reynolds number (Re) using TLBO algorithm. Figure 4 shows the comparison of different algorithms with Reynolds number. Figure 4 shows variation of thermal performance for different number of glass plates at $S = 600$ W/m^2 (using TLBO algorithm). The thermal performance of a flat-plate solar air heater increases with the increase in Reynolds number as seen in Figures 3–5. The thermal efficiency ranges from 31.7385 to 69.9757 with an increasing Reynolds number varying from 2,000 to 20,000 with single glass cover and irradiance of 600 W/m^2 as shown in Table 4. Similarly, the performance range is 38.1378–74.5783 and 43.3532–76.6739 for the same range of Reynolds number and irradiance having two and three glass covers, respectively. The maximum value of thermal efficiency is 76.67% and it is obtained with three glass cover plates and irradiance of 600 W/m^2 at Reynolds number of 20,000. The maximum value of efficiency is obtained at $V = 1.2729$ m/s, tilt angle = 59.5832°, emissivity of plate = 0.8835, ambient temperature = 293.9362 K, and temperature rise = 2.1395 K. Hence, it can be concluded from Tables 4–15 that the thermal performance of a flat-plate solar air heater increases with increase in Reynolds number.

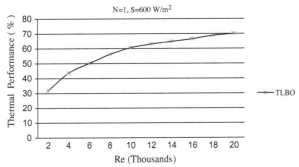

Figure 3. Variation of thermal performance with Reynolds number (Re) (using TLBO algorithm).

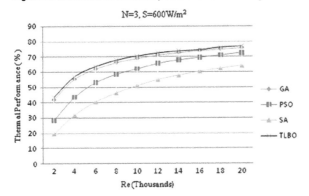

Figure 4. Comparison of thermal performance of different algorithms with Reynolds number.

4.2. Effect of number of glass plates on thermal performance

From Figure 4 and Tables 4–15, it can be seen that as the number of glass plates of solar air heater increases, the thermal efficiency of the solar air heater increases. The thermal performance of the solar air heater is investigated for different sets of glass plates varying from 1 to 3. The range of thermal performance for single glass cover varies from 28.3282% for $I = 1200$ W/m² and $Re = 2,000$ to 69.9757% for $I = 600$ W/m² and $Re = 20,000$. Similarly, for two glass covers, the thermal performance varies from 35.4562% for $I = 1200$ W/m² and $Re = 2,000$ to 74.7183% for $I = 600$ W/m² and $Re = 20,000$, and for three glass covers, the thermal performance varies from 40.2748% for $I = 1200$ W/m² and $Re = 2,000$ to 76.6739% for $I = 600$ W/m² and $Re = 20,000$.

4.3. Effect of solar radiation intensity on thermal performance

Figure 5 shows the effect of solar radiation intensity on thermal performance for $N = 1$. The maximum thermal performance is 76.6739, 76.5913, 76.4188, and 76.0374% for irradiance of 600 W/m², 800 W/m², 1,000 W/m², and 1200 W/m², respectively. Hence, it can be concluded that as the irradiance increases, the thermal performance slightly decreases as can be seen from Figure 5 and Tables 4–15.

5. Conclusions

In the present work, a recently developed optimization algorithm known as TLBO algorithm is used for investigating the thermal performance of a SFPSAH. Maximization of thermal efficiency of SFPSAH is considered as the objective function. The thermal performance is obtained for different Reynolds numbers, irradiance, and number of glass plates. The maximum value of thermal efficiency of 76.67% is obtained with wind velocity of 1.2729 m/s, tilt angle of 59.5832°, plate emissivity of 0.8835, ambient temperature of 3.9362 K, temperature rise of 2.1395 K, irradiance of 600, and Reynolds number of 20,000. The final results obtained by the TLBO algorithm are compared with other optimization algorithms like GA, PSO, and SA and found to be satisfactory. The results also show that the thermal performance increases with the Reynolds number and the number of glass cover plates, but slightly

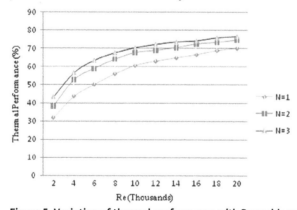

Figure 5. Variation of thermal performance with Reynolds number (Re) for different number of glass plates at $S = 600$ W/m² (using TLBO algorithm).

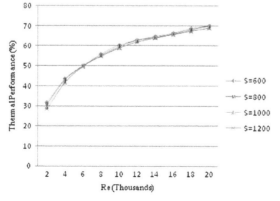

Figure 6. The effect of solar radiation intensity on thermal performance ($N = 1$).

decreases with the increase in irradiance. The TLBO algorithm is an effective algorithm and has potential for finding the optimal set of design and operating parameters at which the thermal performance of a SFPSAH is maximum. The TLBO algorithm may be tried on more complex problems in the near future.

Nomenclature

A_c area of absorber plate (m²)

c_p specific heat of air (J/kg K)

d hydraulic diameter of duct (m)

F_0 heat removal factor referred to outlet temperature (dimensionless)

G mass velocity (kg/sm²)

h convective heat transfer coefficient (W/m²K)

h_w wind convection coefficient (W/m²K)

S irradiance (W/m²)

k index of iteration

k' construction factor

k_{max} maximum number of observations

$\dot{m}$ mass flow rate of air (kg/s)

N number of glass covers (dimensionless)

p_k position of kth iteration

p_r Prandtl number (dimensionless)

Re Reynolds number (dimensionless)

t thickness of insulating material (m)

T_a ambient temperature of air (K)

T_i inlet temperature of air (K)

T_0 outlet temperature of air (K)

T_p temperature of absorber plate (K)

U_0 overall loss coefficient (W/m²K)

U_t top loss coefficient (W/m²K)

v wind velocity (m/s)

x_i experimental value at ith iteration

y_i simulated value at ith iteration

Greek symbols

$(\tau\alpha)$ transmittance–absorptance product (dimensionless)

λ thermal conductivity of air (W/mK)

λ_i thermal conductivity of insulating material (W/mK)

ηth   thermal efficiency (dimensionless)

ϵ_p emissivity of plate (dimensionless)

ϵ_g emissivity of glass cover (dimensionless)

β tilt angle (°)

Funding
The authors received no direct funding for this research.

Author details
R. Venkata Rao[1]
E-mail: ravipudirao@gmail.com
Gajanan Waghmare[1]
E-mail: waghmaregaju@yahoo.com
[1] Department of Mechanical Engineering, S.V. National Institute
  of Technology, Ichchanath, Surat, Gujarat 395 007, India.

References

Ammari, H. D. (2003). A mathematical model of thermal performance of a solar air heater with slats. *Renewable Energy, 28*, 1597–1615. http://dx.doi.org/10.1016/S0960-1481(02)00253-7

ASHRAE Standards. (1997). *Methods of testing to determine the thermal performance of solar collectors.* New York, NY: Author, 93–77.

Basu, M. (2014). Teaching–learning-based optimization algorithm for multi-area economic dispatch. *Energy.* Retrieved from http://dx.doi.org/10.1016/j.energy.2014

Baykasoğlu, A., Hamzadayi, A., & Köse, S. Y. (2014). Testing the performance of teaching–learning based optimization (TLBO) algorithm on combinatorial problems: Flow shop and job shop scheduling cases. *Information Sciences.* Retrieved from http://dx.doi.org/10.1016/j.ins.2014.02.056

Camp, C. V., & Farshchin, M. (2014). Design of space trusses using modified teaching–learning-based optimization. *Engineering Structures, 62–63*, 87–97. http://dx.doi.org/10.1016/j.engstruct.2014.01.020

Chamoli, S., Chauhan, R., Thakur, N. S., & Saini, J. S. (2012). A review of the performance of double pass solar air heater. *Renewable and Sustainable Energy Reviews, 16*, 481–492. http://dx.doi.org/10.1016/j.rser.2011.08.012

Črepinšek, M., Liu, S.-H., & Mernik, L. (2012). A note on teaching–learning-based optimization algorithm. *Information Sciences, 212*, 79–93.

Črepinšek, M., Liu, S.-H., & Mernik, M. (2014). Replication and comparison of computational experiments in applied evolutionary computing: Common pitfalls and guidelines to avoid them. *Applied Soft Computing, 19*, 161–170.

Deb, K. (2000). An efficient constraint handling method for genetic algorithms. *Computer Methods in Applied Mechanics and Engineering, 186*, 311–338. http://dx.doi.org/10.1016/S0045-7825(99)00389-8

El-Sebaii, A. A., Aboul-Enein, S., Ramadan, M. R. I., Shalaby, S. M., & Moharram, B. M. (2011). Investigation of thermal performance of-double pass-flat and v-corrugated plate solar air heaters. *Energy, 36*, 1076–1086. http://dx.doi.org/10.1016/j.energy.2010.11.042

Gill, R. S., Singh, S., & Singh, P. (2012). Low cost solar air heater. *Energy Conversion and Management, 57*, 131–142. http://dx.doi.org/10.1016/j.enconman.2011.12.019

Gupta, D., Solanki, S. C., & Saini, J. S. (1997). Thermohydraulic performance of solar air heaters with roughened absorber plates. *Solar Energy, 61*, 33–42. http://dx.doi.org/10.1016/S0038-092X(97)00005-4

Hegazy, A. A. (1996). Optimization of flow-channel depth for conventional flat plate solar air heaters. *Renewable Energy, 7*, 15–21. http://dx.doi.org/10.1016/0960-1481(95)00117-4

Kalogirou, S. A. (2006). Prediction of flat-plate collector performance parameters using artificial neural networks. *Solar Energy, 80*, 248–259. http://dx.doi.org/10.1016/j.solener.2005.03.003

Karwa, R., & Chitoshiya, G. (2013). Performance study of solar air heater having v-down discrete ribs on absorber plate. *Energy, 55*, 939–955. http://dx.doi.org/10.1016/j.energy.2013.03.068

Klein, S. A. (1975). Calculation of flat plate collector loss coefficients. *Solar Energy, 17*, 79–80. http://dx.doi.org/10.1016/0038-092X(75)90020-1

Lanjewar, A., Bhagoria, J. L., & Sarviya, R. M. (2011). Heat transfer and friction in solar air heater duct with W-shaped rib roughness on absorber plate. *Energy, 36*, 4531–4541. http://dx.doi.org/10.1016/j.energy.2011.03.054

Layek, A., Saini, J. S., & Solanki, S. C. (2007). Second law optimization of a solar air heater having chamfered rib-groove roughness on absorber plate. *Renewable Energy, 32*, 1967–1980. http://dx.doi.org/10.1016/j.renene.2006.11.005

Medina, M. A., Das, S., Coello Coello, C. A., & Ramírez, J. M. (2014). Decomposition-based modern metaheuristic algorithms for multi-objective optimal power flow—A comparative study. *Engineering Applications of Artificial Intelligence, 32*, 10–20. http://dx.doi.org/10.1016/j.engappai.2014.01.016

Mittal, M. K., & Varshney, L. (2006). Optimal thermohydraulic performance of a wire mesh packed solar air heater. *Solar Energy, 80*, 1112–1120. http://dx.doi.org/10.1016/j.solener.2005.10.004

Moghadam, A., & Seifi, A. R. (2014). Fuzzy-TLBO optimal reactive power control variables planning for energy loss minimization. *Energy Conversion and Management, 77*, 208–215. http://dx.doi.org/10.1016/j.enconman.2013.09.036

Niknam, T., Azizipanah-Abarghooee, R., & Rasoul Narimani, M. (2012). A new multi objective optimization approach based on TLBO for location of automatic voltage regulators in distribution systems. *Engineering Applications of Artificial Intelligence, 25*, 1577–1588. http://dx.doi.org/10.1016/j.engappai.2012.07.004

Rao, R. V., Kalyankar, V. D., & Waghmare, G. (2014). Parameters optimization of selected casting processes using teaching–learning based optimization algorithm. *Applied Mathematical Modelling, 38*, 5592–5608. http://dx.doi.org/10.1016/j.apm.2014.04.036

Rao, R. V., & Patel, V. (2012). An elitist teaching–learning-based optimization algorithm for solving complex constrained optimization problems. *International Journal of Industrial Engineering Computations, 3*, 535–560. http://dx.doi.org/10.5267/j.ijiec

Rao, R. V., & Patel, V. (2013). Comparative performance of an elitist teaching–learning-based optimization algorithm for solving unconstrained optimization problems. *International Journal of Industrial Engineering Computations, 4*, 29–50. http://dx.doi.org/10.5267/j.ijiec

Rao, R. V., Savsani, V. J., & Balic, J. (2011). 'Teaching-learning-based optimization algorithm for unconstrained and constrained real parameter optimization problems. *Engineering Optimization, 44*, 1447–1462.

Rao, R. V., Savsani, V. J., & Vakharia, D. P. (2011a). Teaching-learning-based optimization: A novel method for constrained mechanical design optimization problems. *Computer Aided Design, 43*, 303–315. http://dx.doi.org/10.1016/j.cad.2010.12.015

Rao, R. V., Savsani, V. J., & Vakharia, D. P. (2011b). 'Teaching–learning-based optimization: A novel optimization method for continuous non-linear large scale problems. *Information Sciences, 183*(1), 1–15.

Rao, R. V., & Waghmare, G. G. (2014). A comparative study of a teaching–learning-based optimization algorithm on multiobjective unconstrained and constrained functions. *Journal of King Saud University – Computer and Information Sciences, 26*, 332–346.

Satapathy, S. C., & Naik, A. (2014). Modified teaching–learning-based optimization algorithm for global numerical optimization—A comparative study. *Swarm and Evolutionary Computation, 16*, 28–37. http://dx.doi.org/10.1016/j.swevo.2013.12.005

Siddhartha, S., Chauhan, S. R., Varun, V., & Sharma, N. (2012). Thermal performance optimization of smooth flat plate solar air heater (SFPSAH) using simulated annealing: Evaluation and comparisons. *IEEE*, Copyright Notice: 978-1-4673-6008-1/11

Siddhartha, Sharma, N., & Varun (2012). A particle swarm optimization algorithm for optimization of thermal performance of a smooth flat plate solar air heater. *Energy, 38*, 406–413. http://dx.doi.org/10.1016/j.energy.2011.11.026

Sultana, S., & Roy, P. K. (2014). Optimal capacitor placement in radial distribution systems using teaching learning based optimization. *International Journal of Electrical Power & Energy Systems, 54*, 387–398.

Tanda, G. (2011). Performance of solar air heater ducts with different types of ribs on the absorber plate. *Energy, 36,* 6651–6660. http://dx.doi.org/10.1016/j.energy.2011.08.043

Twidell, J., & Weir, T. (2005). *Renewable energy resources* (2nd ed.). Taylor & Francis. ISBN-13:9780419253303

Varun, Sharma, N., Bhat, I. K., & Grover, D. (2011). Optimization of a smooth flat plate solar air heater using stochastic iterative perturbation technique. *Solar Energy, 85,* 2331–2337. http://dx.doi.org/10.1016/j.solener.2011.06.022

Varun, & Siddhartha (2010). Thermal performance optimization of a flat plate solar air heater using genetic algorithm. *Applied Energy, 87,* 1793–1799. http://dx.doi. org/10.1016/j.apenergy.2009.10.015

Veček, N., Mernik, M., & Črepinšek, M. (2014). A chess rating system for evolutionary algorithms: A new method for the comparison and ranking of evolutionary algorithms. *Information Sciences.* Retrieved from http://dx.doi. org/10.1016/j.ins.2014.02.154

Waghmare, G. (2013). Comments on 'a note on teaching-learning-based optimization algorithm'. *Information Sciences, 229,* 159–169.

Yu, K., Wang, X., & Wang, Z. (2014). An improved teaching-learning-based optimization algorithm for numerical and engineering optimization problems. *Journal of Intelligent Manufacturing.* doi:10.1007/s5-014-0918-3

Zou, F., Wang, L., Hei, X., Chen, D., & Yang, D. (2014). Teaching–learning-based optimization with dynamic group strategy for global optimization. *Information Sciences, 273,* 112–131. http://dx.doi.org/ 10.1016/j.ins.2014.03.038

9

Wave propagation in a rotating disc of polygonal cross-section immersed in an inviscid fluid

R. Selvamani[1]*

*Corresponding author: R. Selvamani, Department of Mathematics, Karunya University, Coimbatore, Tamil Nadu 641114, India
E-mail: selvam1729@gmail.com
Reviewing editor: Duc Pham, University of Birmingham, UK

Abstract: In this paper, the wave propagation in a rotating disc of polygonal cross-section immersed in an inviscid fluid is studied using the Fourier expansion collocation method. The equations of motion are derived based on two-dimensional theory of elasticity under the assumption of plane strain-rotating disc of polygonal cross-sections composed of homogeneous isotropic material. The frequency equations are obtained by satisfying the boundary conditions along the irregular surface of the disc using Fourier expansion collocation method. The triangular, square, pentagonal and hexagonal cross-sectional discs are computed numerically for Copper material. Dispersion curves are drawn for non-dimensional wave number and relative frequency shift for longitudinal and flexural (symmetric and anti-symmetric) modes. This work may find applications in navigation and rotating gyroscope .

Subjects: Engineering & Technology; Mathematics & Statistics; Physical Sciences

Keywords: wave propagation in disc; vibration of thermal plate; waves in cylinder; disc immersed in fluid; polygonal cross-section; rotating gyroscope

1. Introduction

The rotating disc of polygonal cross-section is the important structural component in construction of gyroscope to measure the angular velocity of a rotating body. The wave propagation in a disc contact with a fluid finds many applications in the field of structural acoustics, the acoustic microscopic wave interaction in geophysics, characterisation of material properties of thin metal wires, optical fibres and reinforcement filaments used in epoxy, metal and ceramic matrix composites and

ABOUT THE AUTHOR

R. Selvamani is working as an assistant professor in the department of Mathematics in Karunya University, Coimbatore, Tamilnadu, India. He has completed his PhD research from Bharathiar University, Coimbatore. He has published two books which are available in the amazon.com and has published a number of research papers in journals of national and international repute. His research area includes wave propagation in cylinder, panel, plate, composite cylinder and material with arbitrary cross-sections.

PUBLIC INTEREST STATEMENT

This article presents the propagation in a rotating disc of polygonal cross-section immersed in an inviscid fluid using the Fourier expansion collocation method. The computed non-dimensional wave number and relative frequency shift of triangular, square, pentagonal and hexagonal cross-sectional disc are plotted in the form of dispersion curves for Copper material. The results of this study are useful in construction and design of rotating gyroscope, and this method is straightforward. The numerical results for any other polygonal cross-section can be obtained directly for the same frequency equation by substituting geometric values of the boundary of any cross-section analytically or numerically with satisfactory convergence.

non-destructive evaluation of solid structures. The characteristics of wave propagation in rotating disc of polygonal cross-section immersed in fluid have a wide range of applications in the field of machinery, submarine structures, pressure vessel, chemical pipes and metallurgy.

The most general form of harmonic waves in a hollow cylinder of circular cross-section of infinite length has been analysed by Gazis (1959) in two parts. He has presented the frequency equation in part I and numerical results in part II in detailed form. Nagaya (1981, 1983a, 1983b) has discussed wave propagation in a thick-walled pipe, bar of polygonal plate and ring of arbitrary cross-section based on the two-dimensional theory of elasticity. The boundary conditions along both outer and inner free surface of the arbitrary cross-section are satisfied by means of Fourier expansion colloca-tion method. Venkatesan and Ponnusamy (2002) have obtained frequency equation of the free vibration of a solid cylinder of arbitrary cross-section immersed in fluid using the Fourier expansion collocation method. The frequency equations are obtained for longitudinal and flexural vibrations, are studied numerically for elliptic and cardioidal cross-sectional cylinders, and are presented both in tabular and in graphical forms. Later, Ponnusamy (2011) studied the wave propagation in thermoe-lastic plate of arbitrary cross-sections using the Fourier expansion Collocations Method. Ponnusamy and Selvamani (2012, 2013) investigated the dispersion analysis of generalised magneto-thermo-elastic waves in a transversely isotropic cylindrical panel and wave propagation in magneto thermo elastic cylindrical panel, respectively, using Bessel functions. Yazdanpanah Moghadam, Tahani, and Naserian-Nik (2013) obtained an analytical solution of piezolaminated rectangular plate with arbi-trary clamped/simply supported boundary conditions under thermo-electro-mechanical loadings.

Sinha, Plona, Sergio, and Chang (1992) have discussed the axisymmetric wave propagation in a cylindrical shell immersed in fluid, in two parts. In part I, the theoretical analysis of the wave propa-gating modes is discussed and in part II, the axisymmetric modes excluding tensional modes are obtained theoretically and experimentally and are compared. Berliner and Solecki (1996) have stud-ied the wave propagation in fluid-loaded transversely isotropic cylinder. In that paper, part I consists of the analytical formulation of the frequency equation of the coupled system consisting of the cylinder with inner and outer fluid and part II gives the numerical results.

Loy and Lam (1995) discussed the vibration of rotating thin cylindrical panel using Love's first approximation theory. Bhimaraddi (1984) developed a higher order theory for the free vibration analysis of circular cylindrical shell. Zhang (2002) investigated the parametric analysis of frequency of rotating laminated composite cylindrical shell using wave propagation approach. Body wave propagation in rotating thermoelastic media was investigated by Sharma and Grover (2009). The effect of rotation, magneto field, thermal relaxation time and pressure on the wave propagation in a generalised viscoe-lastic medium under the influence of time harmonic source is discussed by Abd-Alla and Bayones (2011). The propagation of waves in conducting piezoelectric solid is studied for the case when the entire medium rotates with a uniform angular velocity by Wauer (1999). Roychoudhuri and Mukhopadhyay (2000) studied the effect of rotation and relaxation times on plane waves in generalised thermo-viscoe-lasticity. Dragomir, Sinnott, Semercigil, and Turan (2014) studied the energy dissipation and critical speed of granular flow in a rotating cylinder and they found that the coefficient of friction has the greatest significance on the centrifuging speed. One-dimensional analysis for magneto-thermo-me-chanical response in a functionally graded annular variable-thickness rotating disc was discussed by Bayat et al. (2014).

In this paper, the wave propagation in rotating disc of polygonal cross-section immersed in an invicid fluid is analysed. The boundary conditions along irregular surfaces have been satisfied by means of Fourier expansion collocation method. The frequency equations of longitudinal and flexural modes are analysed numerically for triangular, square, pentagonal and hexagonal cross-sections, and the computed non-dimensional wave number and relative frequency shift are plotted in graphs.

2. Formulation of the problem

We consider a homogeneous, isotropic rotating elastic disc of polygonal cross-section immersed in an inviscid fluid. The elastic medium is rotating uniformly with an angular velocity $\vec{\Omega} = \Omega\,\vec{n}$, where n is the unit vector in the direction of the axis of rotation. The system displacements and stresses are defined by the polar coordinates r and θ in an arbitrary point inside the disc and denote the displacements u_r in the direction of r and u_θ in the tangential direction θ. The in-plane vibration and displacements of rotating polygonal cross-sectional disc is obtained by assuming that there is no vibration and a displacement along the z axis in the cylindrical coordinate system $(r,\ \theta,\ z)$. The two-dimensional stress equations of motion, strain–displacement relations in the absence of body forces for a linearly elastic medium are

$$\sigma_{rr,r} + r^{-1}\sigma_{r\theta,\theta} + r^{-1}\left(\sigma_{rr} - \sigma_{\theta\theta}\right) + \rho\left(\vec{\Omega}\times(\vec{\Omega}\times\vec{u}) + 2\vec{\Omega}\times\vec{u}_{,t}\right) = \rho u_{r,tt}$$
$$\sigma_{r\theta,r} + r^{-1}\sigma_{\theta\theta,\theta} + 2r^{-1}\sigma_{r\theta} = \rho u_{\theta,tt} \tag{1}$$

where

$$\sigma_{rr} = \lambda\left(e_{rr} + e_{\theta\theta}\right) + 2\mu e_{rr}$$
$$\sigma_{\theta\theta} = \lambda\left(e_{rr} + e_{\theta\theta}\right) + 2\mu e_{\theta\theta}$$
$$\sigma_{r\theta} = 2\mu e_{r\theta} \tag{2}$$

where $\sigma_{rr}, \sigma_{\theta\theta}, \sigma_{r\theta}$ are the stress components, $e_{rr}, e_{\theta\theta}, e_{r\theta}$ are the strain components, ρ is the mass density, Ω is the rotation, t is the time, λ and μ are Lame' constants. The displacement equation of motion has the additional terms with a time-dependent centripetal acceleration $\vec{\Omega}\times(\vec{\Omega}\times\vec{u})$ and $2\vec{\Omega}\times\vec{u}_{,t}$ where $\vec{u} = (u,0,0)$ is the displacement vector and $\vec{\Omega} = (0,\Omega,0)$ is the angular velocity, the comma notation used in the subscript denotes the partial differentiation with respect to the variables.

The strain e_{ij} related to the displacements are given by

$$e_{rr} = u_{r,r}, e_{\theta\theta} = r^{-1}\left(u_r + u_{\theta,\theta}\right), e_{r\theta} = u_{\theta,r} - r^{-1}\left(u_\theta - u_{r,\theta}\right) \tag{3}$$

in which u_r and u_θ is the displacement components along radial and circumferential directions, respectively. The comma in the subscripts denotes the partial differentiation with respect to the variables.

Substituting the Equations 3 and 2 in Equation 1, the following displacement equations of motions are obtained as

$$(\lambda + 2\mu)\left(u_{r,rr} + r^{-1}u_{r,r} - r^{-2}u_r\right) + \mu r^{-2}u_{r,\theta\theta} + r^{-1}(\lambda + \mu)u_{\theta,r\theta} + r^{-2}(\lambda + 3\mu)u_{\theta,\theta} + \rho\Omega^2 u_r = \rho u_{r,tt}$$
$$\tag{4a}$$

$$\mu\left(u_{\theta,rr} + r^{-1}u_{\theta,r} - r^{-2}u_\theta\right) + r^{-2}(\lambda + 2\mu)u_{\theta,\theta\theta} + r^{-2}(\lambda + 3\mu)u_{r,\theta} + r^{-1}(\lambda + \mu)u_{r,r\theta} = \rho u_{\theta,tt} \tag{4b}$$

3. Solutions of solid medium

Equations 4a and 4b are coupled partial differential equation with two displacements components. To uncouple Equations 4a and 4b, we follow Mirsky (1965) by assuming the vibration and displacements along the axial direction z equal to zero. Hence, assuming the solutions of Equations 4a and 4b in the following form

$$u_r(r,\theta,z,t) = \sum_{n=0}^{\infty} \varepsilon_n\left[\left(\phi_{n,r} + r^{-1}\psi_{n,\theta}\right) + \left(\overline{\phi}_{n,r} + r^{-1}\overline{\psi}_{n,\theta}\right)\right]e^{i\omega t} \tag{5a}$$

$$u_\theta(r,\theta,z,t) = \sum_{n=0}^{\infty} \varepsilon_n\left[\left(r^{-1}\phi_{n,\theta} - \psi_{n,r}\right) + \left(r^{-1}\overline{\phi}_{n,\theta} - \overline{\psi}_{n,r}\right)\right]e^{i\omega t} \tag{5b}$$

where $\varepsilon_n = \frac{1}{2}$ for $n = 0$, $\varepsilon_n = 1$ for $n \geq 1$, $i = \sqrt{-1}$, and ω is the angular frequency; $\phi_n(r,\theta)$, $\psi_n(r,\theta)$, $\overline{\phi}_n(r,\theta)$ and $\overline{\psi}_n(r,\theta)$ are the displacement potentials.

By introducing the dimensionless quantities $\overline{\lambda} = (\lambda/\mu)$, $\xi = \sqrt{\rho\omega^2 a^2/\mu}$, $\overline{z} = z/a$, $c^2 = \Omega^2 \xi^2/\omega^2$, $\omega^* = \xi/a\sqrt{\mu/\rho}$, $T = t\sqrt{\mu/\rho}/a$, $x = r/a$, c^2 is the rotational velocity, and substituting Equations 5a and 5b in Equations 4a and 4b, we obtain

$$\left[\left(2+\overline{\lambda}\right)\nabla^2 + \left(\xi^2 + c^2\right)\right]\phi_n = 0 \tag{6}$$

and

$$\left[\nabla^2 + \left(\xi^2 + c^2\right)\right]\psi_n = 0 \tag{7}$$

where $\nabla^2 \equiv \partial^2/\partial x^2 + x^{-1}\partial/\partial x + x^{-2}\partial^2/\partial\theta^2$

From Equation 6, we obtain

$$\left(\nabla^2 + (\alpha a)^2\right)\phi_n = 0 \tag{8}$$

where $(\alpha a)^2 = \left(\xi^2 + c^2\right)/\left(2+\overline{\lambda}\right)$. The solution of Equation 7 for symmetric mode is

$$\phi_n = A_1 J_n\left(\alpha_1 a x\right)\cos n\theta \tag{9}$$

Similarly, the solution for the anti symmetric mode $\overline{\phi}_n$ is obtained by replacing $\cos n\theta$ by $\sin n\theta$ in Equation 9.

$$\overline{\phi}_n = \overline{A}_1 J_n\left(\alpha_1 a x\right)\sin n\theta \tag{10}$$

where J_n is the Bessel function of first kind, $\left(\alpha_1 a\right)^2$ are the roots of Equation 6. Solving Equation 7, we obtain

$$\psi_n = A_2 J_n\left(\alpha_2 a x\right)\sin n\theta \tag{11}$$

for symmetric mode and $\overline{\psi}_n$ is obtained from Equation 11 by replacing $\sin n\theta$ by $\cos n\theta$

$$\overline{\psi}_n = \overline{A}_2 J_n\left(\alpha_2 a x\right)\cos n\theta \tag{12}$$

for the anti symmetric mode. Where $\left(\alpha_2 a\right)^2 = (\xi^2 + c^2)$. If $\left(\alpha_i a\right)^2 < 0$, $i = 1, 2$ then the Bessel function of first kind is to be replaced by the modified Bessel function of the first kind I_n.

4. Solution of fluid medium

In cylindrical coordinates, the acoustic pressure and radial displacement equation of motion for an in viscid fluid are of the form Achenbach (1973)

$$p^f = -B^f\left(u^f{}_{,r} + r^{-1}\left(u^f + v^f{}_{,\theta}\right)\right) \tag{13}$$

and

$$c^{-2}u^f{}_{,tt} = \Delta_{,r} \tag{14}$$

respectively, where (u^f, v^f) is the displacement vector, B^f is the adiabatic bulk modulus, $c = \sqrt{B^f/\rho^f}$ is the acoustic phase velocity of the fluid in which ρ^f is the density of the fluid and

$$\Delta = \left(u^f_{,r} + r^{-1} \left(u^f + v^f_{,\theta} \right) \right) \tag{15}$$

substituting $u^f = \phi^f_{,r}$ and $v^f = r^{-1}\phi^f_{,\theta}$ and seeking the solution of Equation 14 in the form

$$\phi^f(r,\theta,t) = \sum_{n=0}^{\infty} \varepsilon_n \left[\phi^f_n \cos n\theta + \overline{\phi}_n{}^f \sin n\theta \right] e^{i\omega t} \tag{16}$$

where

$$\phi^f_n = A_3 J_n{}^1 \left(\alpha_3 a x \right) \tag{17}$$

for inner fluid. In Equation 17, $\left(\alpha_3 a\right)^2 = \xi^2 / \overline{\rho^f\, B^f}$ in which $\overline{\rho} = \rho/\rho^f$, $\overline{B^f} = B^f/\mu$, J_n^1 is the Bessel function of the first kind and $\overline{\phi}_n^f$ is as same as ϕ^f_n. If $\left(\alpha_3 a\right)^2 < 0$, the Bessel function of first kind is to be replaced by the modified Bessel function of second kind K_n.

By substituting the expression of the displacement vector in terms of ϕ^f and Equations 17 in Equation 13, we could express the acoustic pressure of the disc as

$$p^f = \sum_{n=0}^{\infty} \varepsilon_n A_3 \xi^2 \overline{\rho} J_n^1 \left(\alpha_3 a x \right) \cos n\theta e^{i\omega t} \tag{18}$$

5. Boundary conditions and frequency equations

In this problem, the vibration of a polygonal cross-sectional rotating disc immersed in fluid is considered. Since the boundary is irregular, it is difficult to satisfy the boundary conditions both inner and outer surface of the disc directly. Hence, in the same lines of Nagaya (1983a, 1983b), the Fourier expansion collocation method is applied to satisfy the boundary conditions. Thus, the boundary conditions are obtained as

$$\left(\sigma_{xx} + p^f \right)_l = (\sigma_{xy})_l = (u - u^f)_l = 0 \tag{19}$$

where x is the coordinate normal to the boundary and y is the tangential coordinate to the boundary, σ_{xx} is the normal stress, σ_{xy} is the shearing stress and $(\ )_l$ represents the value at the lth segment of the boundary. The first and last conditions in Equation 19 are due to the continuity of the stresses and displacements of the disc and fluid on the curved surface. If the angle γ_l between the normal to the segment and the reference axis is assumed to be constant, the transformed expression for the stresses are given by

$$\sigma_{xx} = \lambda \left(u_{,r} + r^{-1} \left(u + v_{,\theta} \right) \right) + 2\mu \left[u_{,r} \cos^2 \left(\theta - \gamma_l \right) + r^{-1} \left(u + v_{,\theta} \right) \sin^2 \left(\theta - \gamma_l \right) \right.$$
$$\left. + 0.5 \left(r^{-1} \left(v - u_{,\theta} \right) - v_{,r} \right) \sin 2 \left(\theta - \gamma_l \right) \right]$$
$$\sigma_{xy} = \mu [u_{,r} - r^{-1} \left(v_{,\theta} + u \right) \sin 2 \left(\theta - \gamma_l \right) + \left(r^{-1} \left(u_{,\theta} - v \right) + v_{,r} \right) \sin 2 \left(\theta - \gamma_l \right)] \tag{20}$$

The boundary conditions in Equation 19 are transformed for rotating disc as

$$\left[(S_{xx})_l + \left(\overline{S}_{xx} \right)_l \right] e^{i\left(\varepsilon \overline{z} + \Omega T \right)} = 0$$
$$\left[(S_{xy})_l + \left(\overline{S}_{xy} \right)_l \right] e^{i\left(\varepsilon \overline{z} + \Omega T \right)} = 0$$
$$\left[(S_r)_l + \left(\overline{S}_r \right)_l \right] e^{i\left(\varepsilon \overline{z} + \Omega T \right)} = 0 \tag{21}$$

where

$$S_{xx} = 0.5 \left(e_0^1 A_{10} + e_0^2 A_{20} \right) + \sum_{n=1}^{\infty} \left(e_n^1 A_{1n} + e_n^2 A_{2n} + e_n^3 A_{3n} \right)$$
$$S_{xy} = 0.5 \left(f_0^1 A_{10} + f_0^2 A_{20} \right) + \sum_{n=1}^{\infty} \left(f_n^1 A_{1n} + f_n^2 A_{2n} + f_n^3 A_{3n} \right)$$
$$S_{t} = 0.5 \left(g_0^1 A_{10} + g_0^2 A_{20} \right) + \sum_{n=1}^{\infty} \left(g_n^1 A_{1n} + g_n^2 A_{2n} + g_n^3 A_{3n} \right) \tag{22}$$

$$\bar{S}_{xx} = 0.5 \left(\bar{e}_0^3 \bar{A}_{30} \right) + \sum_{n=1}^{\infty} \left(\bar{e}_n^1 \bar{A}_{1n} + \bar{e}_n^2 \bar{A}_{2n} + \bar{e}_n^3 \bar{A}_{3n} \right)$$
$$\bar{S}_{xy} = 0.5 \left(\bar{f}_0^3 \bar{A}_{30} \right) + \sum_{n=1}^{\infty} \left(\bar{f}_n^1 \bar{A}_{1n} + \bar{f}_n^2 \bar{A}_{2n} + \bar{f}_n^3 \bar{A}_{3n} \right)$$
$$\bar{S}_{t} = 0.5 \left(\bar{g}_0^3 \bar{A}_{30} \right) + \sum_{n=1}^{\infty} \left(\bar{g}_n^1 \bar{A}_{1n} + \bar{g}_n^2 \bar{A}_{2n} + \bar{g}_n^3 \bar{A}_{3n} \right) \tag{23}$$

The coefficients for $e_n^j - \bar{g}_n^j$ are given in Appendix A.

Performing the Fourier series expansion to Equation 19 along the boundary, the boundary conditions are expanded in the form of double Fourier series. In the symmetric mode, the boundary conditions are obtained as

$$\sum_{m=0}^{\infty} \varepsilon_m \left[E_{m0}^1 A_{10} + E_{m0}^2 A_{20} + \sum_{n=1}^{\infty} \left(E_{mn}^1 A_{1n} + E_{mn}^2 A_{2n} + E_{mn}^3 A_{3n} \right) \right] \cos m\theta = 0$$
$$\sum_{m=1}^{\infty} \left[F_{m0}^1 A_{10} + F_{m0}^2 A_{20} + \sum_{n=1}^{\infty} \left(F_{mn}^1 A_{1n} + F_{mn}^2 A_{2n} + F_{mn}^3 A_{3n} \right) \right] \sin m\theta = 0$$
$$\sum_{m=0}^{\infty} \varepsilon_m \left[G_{m0}^1 A_{10} + G_{m0}^2 A_{20} + \sum_{n=1}^{\infty} \left(G_{mn}^1 A_{1n} + G_{mn}^2 A_{2n} + G_{mn}^3 A_{3n} \right) \right] \cos m\theta = 0 \tag{24}$$

and for antisymmetric mode, the boundary conditions are expresses as

$$\sum_{m=1}^{\infty} \left[\bar{E}_{m0}^3 \bar{A}_{30} + \sum_{n=1}^{\infty} \left(\bar{E}_{mn}^1 \bar{A}_{1n} + \bar{E}_{mn}^2 \bar{A}_{2n} + \bar{E}_{mn}^3 \bar{A}_{3n} \right) \right] \sin m\theta = 0$$
$$\sum_{m=0}^{\infty} \varepsilon_m \left[\bar{F}_{m0}^3 \bar{A}_{30} + \sum_{n=1}^{\infty} \left(\bar{F}_{mn}^1 \bar{A}_{1n} + \bar{F}_{mn}^2 \bar{A}_{2n} + \bar{F}_{mn}^3 \bar{A}_{3n} \right) \right] \cos m\theta = 0$$
$$\sum_{m=1}^{\infty} \left[\bar{G}_{m0}^3 \bar{A}_{30} + \sum_{n=1}^{\infty} \left(\bar{G}_{mn}^1 \bar{A}_{1n} + \bar{G}_{mn}^2 \bar{A}_{2n} + \bar{G}_{mn}^3 \bar{A}_{3n} \right) \right] \sin m\theta = 0 \tag{25}$$

where

$$E_{mn}^j = \left(2\varepsilon_n / \pi \right) \sum_{i=1}^{I} \int_{\theta_{i-1}}^{\theta_i} e_n^j \left(R_i, \theta \right) \cos m\theta \, d\theta$$
$$F_{mn}^j = \left(2\varepsilon_n / \pi \right) \sum_{i=1}^{I} \int_{\theta_{i-1}}^{\theta_i} f_n^j \left(R_i, \theta \right) \sin m\theta \, d\theta$$
$$G_{mn}^j = \left(2\varepsilon_n / \pi \right) \sum_{i=1}^{I} \int_{\theta_{i-1}}^{\theta_i} g_n^j \left(R_i, \theta \right) \cos m\theta \, d\theta \tag{26}$$

$$\bar{E}_{mn}^j = \left(2\varepsilon_n / \pi \right) \sum_{i=1}^{I} \int_{\theta_{i-1}}^{\theta_i} \bar{e}_n^j \left(R_i, \theta \right) \sin m\theta \, d\theta$$
$$\bar{F}_{mn}^j = \left(2\varepsilon_n / \pi \right) \sum_{i=1}^{I} \int_{\theta_{i-1}}^{\theta_i} \bar{f}_n^j \left(R_i, \theta \right) \cos m\theta \, d\theta$$
$$\bar{G}_{mn}^j = \left(2\varepsilon_n / \pi \right) \sum_{i=1}^{I} \int_{\theta_{i-1}}^{\theta_i} \bar{g}_n^j \left(R_i, \theta \right) \sin m\theta \, d\theta \tag{27}$$

and where $j = 1, 2,$ and 3, I is the number of segments, R_i is the coordinate r at the boundary and N is the number of truncation of the Fourier series. The frequency equations are obtained by truncating the series to $N + 1$ terms, and equating the determinant of the coefficients of the amplitude $A_{in} = 0$ and $\overline{A}_{in} = 0$, for symmetric and antisymmetric modes of vibrations. Thus, the frequency equation for the symmetric mode is obtained from Equation 24, by equating the determinant of the coefficient matrix of $A_{in} = 0$. Therefore, we have

$$
\begin{vmatrix}
E^1_{00} & E^2_{00} & E^1_{01} & \cdots & E^1_{0N} & E^2_{01} & \cdots & E^2_{0N} & E^3_{01} & \cdots & E^3_{0N} \\
\vdots & \vdots & \vdots & & \vdots & \vdots & & \vdots & \vdots & & \vdots \\
E^1_{N0} & E^2_{N0} & E^1_{N1} & \cdots & E^1_{NN} & E^2_{N1} & \cdots & E^2_{NN} & E^3_{N1} & \cdots & E^3_{NN} \\
F^1_{00} & F^2_{00} & F^1_{01} & \cdots & F^1_{0N} & F^2_{01} & \cdots & F^2_{0N} & F^3_{01} & \cdots & F^3_{0N} \\
\vdots & \vdots & \vdots & & \vdots & \vdots & & \vdots & \vdots & & \vdots \\
F^1_{N0} & F^2_{N0} & F^1_{N1} & \cdots & F^1_{NN} & F^2_{N1} & \cdots & F^2_{NN} & F^3_{N1} & \cdots & F^3_{NN} \\
G^1_{00} & G^2_{00} & G^1_{01} & \cdots & G^1_{0N} & G^2_{01} & \cdots & G^2_{0N} & G^3_{01} & \cdots & G^3_{0N} \\
\vdots & \vdots & \vdots & & \vdots & \vdots & & \vdots & \vdots & & \vdots \\
G^1_{N0} & G^2_{N0} & G^1_{N1} & \cdots & G^1_{NN} & G^2_{N1} & \cdots & G^2_{NN} & g^3_{N1} & \cdots & G^3_{NN}
\end{vmatrix} = 0
$$

(28)

Similarly, the frequency equation for antisymmetric mode is obtained from the Equation 25 by equating the determinant of the coefficient matrix of $\overline{A}_{in}$ to zero. Therefore, for the antisymmetric mode, the frequency equation obtained as

$$
\begin{vmatrix}
\overline{E}^3_{10} & \overline{E}^1_{11} & \cdots & \overline{E}^1_{1N} & \overline{E}^2_{11} & \cdots & \overline{E}^2_{1N} & \overline{E}^3_{11} & \cdots & \overline{E}^3_{1N} \\
\vdots & \vdots & & \vdots & \vdots & & \vdots & \vdots & & \vdots \\
\overline{E}^3_{N0} & \overline{E}^1_{N1} & \cdots & \overline{E}^1_{NN} & \overline{E}^2_{N1} & \cdots & \overline{E}^2_{NN} & \overline{E}^3_{N1} & \cdots & \overline{E}^3_{NN} \\
\overline{F}^3_{10} & \overline{F}^1_{11} & \cdots & \overline{F}^1_{1N} & \overline{F}^2_{11} & \cdots & \overline{F}^2_{1N} & \overline{F}^3_{11} & \cdots & \overline{F}^3_{1N} \\
\vdots & \vdots & & \vdots & \vdots & & \vdots & \vdots & & \vdots \\
\overline{F}^3_{N0} & \overline{F}^1_{N1} & \cdots & \overline{F}^1_{NN} & \overline{F}^2_{N1} & \cdots & \overline{F}^2_{NN} & \overline{F}^3_{N1} & \cdots & \overline{F}^3_{NN} \\
\overline{G}^3_{10} & \overline{G}^1_{11} & \cdots & \overline{G}^1_{1N} & \overline{G}^2_{11} & \cdots & \overline{G}^2_{1N} & \overline{G}^3_{11} & \cdots & \overline{G}^3_{1N} \\
\vdots & \vdots & & \vdots & \vdots & & \vdots & \vdots & & \vdots \\
\overline{G}^3_{N0} & \overline{G}^1_{N1} & \cdots & \overline{G}^1_{NN} & \overline{G}^2_{N1} & \cdots & \overline{G}^2_{NN} & \overline{G}^3_{N1} & \cdots & \overline{G}^3_{NN}
\end{vmatrix} = 0
$$

(29)

6. Polygonal cross-sectional disc with fluid and without rotation

The wave propagation in a polygonal cross-sectional disc is obtained by substituting the rotation speed $\Omega = 0$ in the corresponding equations and expressions in the previous section, the problem is reduced to wave propagation in polygonal cross-sectional disc immersed in fluid. Substituting Equations 2, 3 in Equation 1 along with $\Omega = 0$, we get the displacement equations as follows:

$$
(\lambda + 2\mu)\left(u_{r,rr} + r^{-1}u_{r,r} - r^{-2}u_r\right) + \mu r^{-2}u_{r,\theta\theta} + r^{-1}(\lambda + \mu)u_{\theta,r\theta}
$$
$$
+ r^{-2}(\lambda + 3\mu)u_{\theta,\theta} = \rho u_{r,tt}
$$

(30a)

$$
\mu\left(u_{\theta,rr} + r^{-1}u_{\theta,r} - r^{-2}u_\theta\right) + r^{-2}(\lambda + 2\mu)u_{\theta,\theta\theta} + r^{-2}(\lambda + 3\mu)u_{r,\theta}
$$
$$
+ r^{-1}(\lambda + \mu)u_{r,r\theta} = \rho u_{\theta,tt}
$$

(30b)

Equations 30a and 30b are coupled partial differential equation with two displacements components. To uncouple Equations 30a and 30b, use Equation 4 in Equations 30a and 30b, we obtain a differential equations of the form

$$\left[\left(2+\bar{\lambda}\right)\nabla^2+\xi^2\right]\phi_n'=0 \tag{31}$$

and

$$\left[\nabla^2+\xi^2\right]\psi_n'=0 \tag{32}$$

in which $\nabla^2 \equiv \partial^2\big/\partial x^2 + x^{-1}\partial/\partial x + x^{-2}\partial^2\big/\partial\theta^2$

From Equation 6, we obtain

$$\left(\nabla^2+(\beta a)^2\right)\phi_n=0 \tag{33}$$

where $(\beta a)^2 = \xi^2\big/\left(2+\bar{\lambda}\right)$. The solution of Equation 31 for symmetric mode is

$$\phi_n' = A_1'J_n\left(\beta_1 ax\right)\cos n\theta \tag{34}$$

Similarly the solution for the anti symmetric mode $\overline{\phi}_n'$ is obtained by replacing $\cos n\theta$ by $\sin n\theta$ in Equation 34.

$$\overline{\phi}_n' = \overline{A}_1'J_n\left(\beta_1 ax\right)\sin n\theta \tag{35}$$

where J_n is the Bessel function of first kind, $\left(\alpha_1 a\right)^2$ are the roots of Equation 6. Solving Equation 7, we obtain

$$\psi_n' = A_2'J_n\left(\beta_2 ax\right)\sin n\theta \tag{36}$$

for symmetric mode and $\overline{\psi}_n'$ is obtained from Equation 36 by replacing $\sin n\theta$ by $\cos n\theta$

$$\overline{\psi}_n' = \overline{A}_2'J_n\left(\beta_2 ax\right)\cos n\theta \tag{37}$$

for the antisymmetric mode. Where $\left(\beta_2 a\right)^2 = \xi^2$. If $\left(\beta_i a\right)^2 < 0, i = 1, 2$ then the Bessel function of first kind is to be replaced by the modified Bessel function of the first kind I_n.

By applying the same procedure as discussed the previous sections, and using the boundary conditions given in Equation 19, we obtain the frequency equations for polygonal cross-sectional disc immersed in fluid as

$$\begin{vmatrix} b_{11} & b_{12} & b_{13} \\ b_{21} & b_{22} & b_{23} \\ b_{31} & b_{32} & b_{33} \end{vmatrix} = 0 \tag{38}$$

where

$$b_{11} = 2\left\{n(n-1)J_n\left(\beta_1 ax\right) + \left(\beta_1 ax\right)J_{n+1}\left(\beta_1 ax\right)\right\}\cos 2\left(\theta-\gamma_l\right)\cos n\theta$$
$$-x^2\left\{\left(\beta_1 a\right)^2\left(\bar{\lambda}+2\cos^2\left(\theta-\gamma_l\right)\right)\right\}J_n\left(\beta_1 ax\right)\cos n\theta$$

$$b_{12} = 2n\left\{(n-1)J_n\left(\beta_2 ax\right) + \left(\beta_2 ax\right)J_{n+1}\left(\beta_2 ax\right)\right\}\cos 2\left(\theta-\gamma_l\right)\cos n\theta$$
$$+2\left\{\left(n(n-1)-\left(\beta_2 ax\right)^2\right)J_n\left(\beta_2 ax\right) + \left(\beta_2 ax\right)J_{n+1}\left(\beta_2 ax\right)\right\}\sin 2\left(\theta-\gamma_l\right)\sin n\theta$$

$$b_{13} = \xi^2\overline{\rho}J^1{}_n\left(\beta_3 ax\right)\cos n\theta$$

$$b_{21} = 2\left\{n(n-1)-\left(\beta_1 ax\right)^2J_n\left(\beta_1 ax\right) + \left(\beta_1 ax\right)J_{n+1}\left(\beta_1 ax\right)\right\}\sin 2\left(\theta-\gamma_l\right)\cos n\theta$$
$$+2n\left\{\left(\beta_1 ax\right)J_{n+1}\left(\beta_1 ax\right) - (n-1)J_n\left(\beta_1 ax\right)\right\}\cos 2\left(\theta-\gamma_l\right)\sin n\theta$$

$$b_{22} = 2n \left\{ (n-1) J_n \left(\beta_2 ax \right) - \left(\beta_2 ax \right) J_{n+1} \left(\beta_2 ax \right) \right\} \sin 2 \left(\theta - \gamma_l \right) \cos n\theta$$
$$+ 2 \left\{ \left(\beta_2 ax \right) J_{n+1} \left(\beta_2 ax \right) - \left(n(n-1) - \left(\beta_2 ax \right)^2 \right) J_n \left(\beta_2 ax \right) \right\} \cos 2 \left(\theta - \gamma_l \right) \sin n\theta$$

$$b_{33} = 0 \tag{39}$$

$$b_{31} = \left\{ n J_n \left(\beta_1 ax \right) - \left(\beta_1 ax \right) J_{n+1} \left(\beta_1 ax \right) \right\} \cos n\theta$$

$$b_{32} = n J_n \left(\beta_2 ax \right) \cos n\theta$$

$$b_{33} = - \left[n J^1_n \left(\beta_3 ax \right) - \left(\beta_3 ax \right) J^1_{n+1} \left(\beta_3 ax \right) \right] \cos n\theta$$

7. Polygonal cross-sectional disc without fluid and rotation

The free vibration of homogeneous isotropic disc of polygonal cross-section without fluid medium and without rotation can be recovered from the present analysis by omitting the fluid medium and setting $\Omega = 0$, along win the corresponding equations and solutions in the previous sections, then the problem of rotating disc of polygonal cross-section immersed in fluid is converted into a problem of a two-dimensional vibration analysis of polygonal cross-sectional disc. The frequency equations of an polygonal cross-sectional disc without fluid and rotation is obtained as

Stress-free (Unclamped edge), which leads to

$$\left(\sigma_{xx} \right)_i = \left(\sigma_{xy} \right)_i = 0 \tag{40}$$

Rigidly fixed (Clamped edge), implies that

$$\left(u_r \right)_i = \left(u_\theta \right)_i = 0 \tag{41}$$

By applying the same procedure as discussed in the previous section, the stresses are transformed for a disc without fluid and rotation is given as follows.

$$\sigma'_{xx} = \lambda \left(u_{,r} + r^{-1} \left(u + v_{,\theta} \right) \right) + 2\mu [u_{,r} \cos^2 \left(\theta - \gamma_i \right) + r^{-1} \left(u + v_{,\theta} \right) \sin^2 \left(\theta - \gamma_i \right)$$
$$+ 0.5 \left(r^{-1} \left(v - u_{,\theta} \right) - v_{,r} \right) \sin 2 \left(\theta - \gamma_i \right)]$$

$$\sigma'_{xy} = \mu [u_{,r} - r^{-1} \left(v_{,\theta} + u \right) \sin 2 \left(\theta - \gamma_i \right) + \left(r^{-1} \left(u_{,\theta} - v \right) + v_{,r} \right) \sin 2 \left(\theta - \gamma_i \right)] \tag{42}$$

Substituting Equations 9–12 in Equation 40, the boundary conditions are transformed for stress-free disc as follows:

$$\left[\left(S'_{xx} \right)_i + \left(\overline{S}'_{xx} \right)_i \right] e^{i\Omega T_a} = 0$$

$$\left[\left(S'_{xy} \right)_i + \left(\overline{S}'_{xy} \right)_i \right] e^{i\Omega T_a} = 0 \tag{43}$$

where

$$S'_{xx} = 0.5 \left(k_0^1 P_{10} \right) + \sum_{n=1}^{\infty} \left(k_n^1 P_{1n} + k_n^2 P_{2n} \right), S'_{xy} = 0.5 \left(l_0^1 Q_{10} \right) + \sum_{n=1}^{\infty} \left(l_n^1 Q_{1n} + l_n^2 Q_{2n} \right) \tag{44}$$

$$\overline{S}_{xx} = 0.5 \left(\overline{k}_0^3 \overline{P}_{30} \right) + \sum_{n=1}^{\infty} \left(\overline{k}_n^1 \overline{P}_{1n} + \overline{k}_n^2 \overline{P}_{2n} \right), \overline{S}'_{xy} = 0.5 \left(\overline{l}_0^3 \overline{Q}_{30} \right) + \sum_{n=1}^{\infty} \left(\overline{l}_n^1 \overline{Q}_{1n} + \overline{l}_n^2 \overline{A}_{2n} \right) \tag{45}$$

in which

$$k_n^1 = 2\left\{n(n-1)J_n(\alpha_1 ax) + (\alpha_1 ax)J_{n+1}(\alpha_1 ax)\right\}\cos 2(\theta-\gamma_i)\cos n\theta$$
$$-x^2\left\{(\alpha_1 a)^2\left(\bar{\lambda}+2\cos^2(\theta-\gamma_i)\right)\right\}J_n(\alpha_1 ax)\cos n\theta$$

$$k_n^2 = 2n\left\{(n-1)J_n(\alpha_2 ax) + (\alpha_2 ax)J_{n+1}(\alpha_2 ax)\right\}\cos 2(\theta-\gamma_i)\cos n\theta$$
$$+2\left\{\left(n(n-1)-(\alpha_2 ax)^2\right)J_n(\alpha_2 ax) + (\alpha_2 ax)J_{n+1}(\alpha_2 ax)\right\}\sin 2(\theta-\gamma_i)\sin n\theta$$

$$l_n^1 = \left\{[2n(n-1)-(\alpha_1 ax)^2]J_n(\alpha_1 ax) + 2(\alpha_1 ax)J_{n+1}(\alpha_1 ax)\right\}\sin 2(\theta-\gamma_i)\cos n\theta$$
$$+2n\left\{(\alpha_1 ax)J_{n+1}(\alpha_1 ax) - (n-1)J_n(\alpha_1 ax)\right\}\cos 2(\theta-\gamma_i)\sin n\theta$$

$$l_n^2 = 2n\left\{(n-1)J_n(\alpha_2 ax) - (\alpha_2 ax)J_{n+1}(\alpha_2 ax)\right\}\sin 2(\theta-\gamma_i)\cos n\theta$$
$$+2\left\{(\alpha_2 ax)J_{n+1}(\alpha_2 ax) - \left(n(n-1)-(\alpha_2 ax)^2\right)J_n(\alpha_2 ax)\right\}\cos 2(\theta-\gamma_i)\sin n\theta \qquad (46)$$

The barred expressions for the antisymmetric case are obtained by replacing $\cos n\theta$ by $\sin n\theta$ and $\sin n\theta$ by $\cos n\theta$ in Equation 46. Performing the Fourier series expansion to Equation 40 and along the boundary, the boundary conditions along the surfaces are expanded in the form of double Fourier series. In the symmetric mode, the boundary conditions obtained as

$$\sum_{m=0}^{\infty}\varepsilon_m\left[\hat{K}_{m0}^1 P_{10} + \sum_{n=1}^{\infty}\left(\hat{K}_{mn}^1 P_{1n}+\hat{K}_{mn}^2 P_{2n}\right)\right]\cos m\theta = 0 \qquad (47a)$$

$$\sum_{m=1}^{\infty}\left[\hat{L}_{m0}^2 P_{10}\sum_{n=1}^{\infty}\left(\hat{L}_{mn}^1 P_{1n}+\hat{L}_{mn}^2 P_{2n}\right)\right]\sin m\theta = 0 \qquad (47b)$$

where

$$\hat{K}_{mn}^j = (2\varepsilon_n/\pi)\sum_{i=1}^{I}\int_{\theta_{i-1}}^{\theta_i} k_n^j\left(\hat{R}_i,\theta\right)\cos m\theta\, d\theta,\ \hat{L}_{mn}^j = (2\varepsilon_n/\pi)\sum_{i=1}^{I}\int_{\theta_{i-1}}^{\theta_i} l_n^j\left(\hat{R}_i,\theta\right)\sin m\theta\, d\theta \qquad (48)$$

where $j = 1, 2$ I the number of segments is, $\hat{R}_i$ is the coordinate $\hat{r}$ at the boundary, N is the number of truncation of the Fourier series. For the non-trivial solution of the systems of equations given in Equations 47a and 47b, the determinant of the coefficient matrix $P_{in} = 0$ and these determinants give the frequencies of symmetric mode of vibration. Therefore, we have

$$\begin{vmatrix} \hat{K}_{00}^1 & \hat{K}_{01}^1 & \cdots & \hat{K}_{0N}^1 & \hat{K}_{01}^2 & \cdots & \hat{K}_{0N}^2 \\ \vdots & \vdots & & \vdots & \vdots & & \vdots \\ \hat{K}_{N0}^1 & \hat{K}_{N1}^1 & \cdots & \hat{K}_{NN}^1 & \hat{K}_{N1}^2 & \cdots & \hat{K}_{NN}^2 \\ \hat{L}_{00}^1 & \hat{L}_{01}^1 & \cdots & \hat{L}_{0N}^1 & \hat{L}_{01}^2 & \cdots & \hat{L}_{0N}^2 \\ \vdots & \vdots & & \vdots & \vdots & & \vdots \\ \hat{L}_{N0}^1 & \hat{L}_{N1}^1 & \cdots & \hat{L}_{NN}^1 & \hat{L}_{N1}^2 & \cdots & \hat{L}_{NN}^2 \end{vmatrix} = 0 \qquad (49)$$

Similarly, the frequency equations for the antisymmetric mode are obtained by replacing $\cos n\theta$ by $\sin\theta$ and $\sin\theta$ by $\cos n\theta$ in the above corresponding equations.

8. Relative frequency shift

Relative frequency shift plays an important role in construction of rotating gyroscope, acoustic sensors and actuators. The frequency shift of the wave due to rotation is defined as $\Delta\omega = \omega(\Omega) - \omega(0)$. Ω being the angular rotation; the relative frequency shift is given by

128   Mechanical Engineering: Design, Processes and Systems

$$R.F.S = \left| \frac{\Delta\omega}{\omega} \right| = \left| \frac{\omega(\Omega) - \omega(0)}{\omega(0)} \right| \qquad\qquad (50)$$

where $\omega(0)$ is the frequency of the waves in the absence of rotation.

9. Numerical results and discussions

The frequency equations are obtained in symmetric and antisymmetric cases given in Equations 26 and 27 are analysed numerically for rotating disc of triangular, square, pentagonal and hexagonal cross-sections. The material properties used for the computation are as follows: for the solid the Poisson ratio $v = 0.3$, density $\rho = 7{,}849$ kg/m³ and the Young's modulus $E = 2.139 \times 10^{11}$ N/m² and for the fluid: the density $\rho^f = 1{,}000$ Kg/m³ and the phase velocity $c = 1{,}500$ m/s. The geometric relations for the polygonal cross-section are taken from Equation 19 of Nagaya (1981) as

$$R_i/b = \left[\cos\left(\theta - \gamma_i\right) \right]^{-1}$$

where b is the apothem and these relation is used directly for the calculation. The dimensionless frequency which are complex in nature, are computed by fixing the wave number for different thickness using Secant method for polygonal disc immersed in fluid. The polygonal cross-sectional disc in the range $\theta = 0$ and $\theta = \pi$ is divided into many segments for convergence of frequency in such a way that the distance between any two segments is negligible. Integration is performed for each segment numerically by use of Gauss five point formula.

In Tables 1 and 2, a comparison is made for the frequency amongst triangular, square, pentagon and hexagon cross-sectional disc with fluid and without rotation for different aspect ratios. From these Tables it is clear that, as the vibration modes increases the non-dimensional frequencies are also increases in all the four cross-sectional discs and the non-dimensional frequency profiles exhibit high energy in hexagonal disc for increasing aspect ratio. Another spectrum of frequencies of triangular, square, pentagon and hexagon discs with the increasing aspect ratio in the absence of fluid medium and rotational speed is discussed in Tables 3 and 4. From Tables 3 and 4, it is observed

Table 1. The non-dimensional frequency of triangular and square cross-sectional disc with fluid and without rotation for different aspect ratios

Mode	Triangle			Square		
	a/b = 1.0	a/b = 1.5	a/b = 2.0	a/b = 1.0	a/b = 1.5	a/b = 2.0
S1	3.5251	3.6403	3.4426	3.2076	3.0775	3.2511
S2	4.8474	4.9872	4.8692	4.7112	4.8430	4.9522
S3	5.5647	5.6966	5.6625	5.4685	5.4844	5.8029
S4	6.2604	6.2625	6.3683	6.1122	6.1911	6.3618
S5	6.9675	6.9685	7.0676	6.8276	6.8976	7.0225

Table 2. The non-dimensional frequency of pentagon and hexagon cross-sectional disc with fluid and without rotation for different aspect ratios

Mode	Pentagon			Hexagon		
	a/b = 1.0	a/b = 1.5	a/b = 2.0	a/b = 1.0	a/b = 1.5	a/b = 2.0
S1	4.2469	4.1053	4.2473	4.2469	4.2426	4.3841
S2	4.9545	4.9501	5.0964	4.9511	4.9497	5.0927
S3	5.6625	5.6556	5.8041	5.6624	5.6593	5.8041
S4	6.3693	6.2288	6.3712	6.3668	6.2257	6.5118
S5	7.0732	6.9290	7.2125	7.0680	6.9366	7.0704

Table 3. The non-dimensional frequency of triangular and square cross-sectional disc without fluid and without rotation for different aspect ratios

Mode	Triangle			Square		
	$a/b = 1.0$	$a/b = 1.5$	$a/b = 2.0$	$a/b = 1.0$	$a/b = 1.5$	$a/b = 2.0$
A1	4.1440	4.1536	4.2475	4.0336	4.0978	4.2434
A2	4.8602	4.8621	4.9428	4.7134	4.8007	4.9526
A3	5.5753	5.5712	5.6615	5.4087	5.5055	5.6599
A4	6.2691	6.2785	6.3704	6.1590	6.2154	6.3711
A5	6.9696	6.9857	7.0715	6.8707	6.9220	7.0726

Table 4. The non-dimensional frequency of pentagon and hexagon cross-sectional disc without fluid and without rotation for different aspect ratios

Mode	Pentagon			Hexagon		
	$a/b = 1.0$	$a/b = 1.5$	$a/b = 2.0$	$a/b = 1.0$	$a/b = 1.5$	$a/b = 2.0$
A1	4.2377	4.1785	4.3766	4.2480	4.2232	4.3144
A2	4.9598	4.8095	5.0695	4.9535	4.8513	5.2585
A3	5.6645	5.5171	5.8027	5.6723	5.5509	5.8042
A4	6.3672	6.2246	6.3815	6.5083	6.1846	6.5117
A5	7.0761	6.9380	7.2124	7.0833	7.0781	7.2201

that as the modes are increases the non-dimensional frequency also increases, whereas the values in the hexagonal modes are almost exponentially increasing with increasing aspect ratio. The amplitude of the all modes of vibrations exhibits high energy in the absence of fluid and rotational parameter (Figure 1).

9.1. Triangular and pentagonal cross-sections
In triangular and pentagonal cross-sectional discs, the vibrational displacements are symmetrical about the x axis for the longitudinal mode and anti-symmetrical about the y axis for the flexural mode since the cross-section is symmetric about only one axis. Therefore, n and m are chosen as 0, 1, 2, 3, … in Equation 24 for longitudinal mode and n, m = 1, 2, 3, … in Equation 25 for the flexural mode.

9.2. Square and hexagonal cross-sections
In the case of longitudinal vibration of square and hexagonal cross-sectional disc, the displacements are symmetrical about both major and minor axes since both the cross-sections are symmetrical about both the axes. Therefore, the frequency equation is obtained by choosing both terms of n and m is chosen as 0, 2, 4, 6, … in Equation 24. During flexural motion, the displacements are antisymmetrical about the major axis and symmetrical about the minor axis. Hence, the frequency equation is obtained by choosing n, m = 1, 3, 5, … in Equation 25.

9.3. Dispersion curves
The notations used in the Figures namely, Lm, FSm and FASm denote the longitudinal, flexural symmetric and flexural antisymmetric modes of vibrations.

The non-dimensional wave number is calculated with increasing non-dimensional frequency for the longitudinal and flexural (symmetric and antisymmetric) modes of triangle, square, pentagon and hexagon cross-sectional discs with different values of the rotational speed Ω = 0.2, 0.4 in Figures 2–5. From these curves, it can be seen that the frequencies are increases in all the three kinds of vibrational modes such as longitudinal and flexural (symmetric and antisymmetric) with

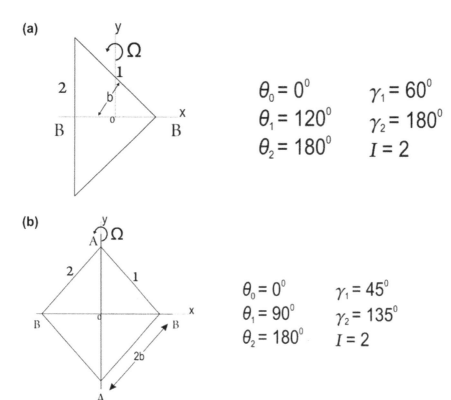

(a)

$$\theta_0 = 0^0 \qquad \gamma_1 = 60^0$$
$$\theta_1 = 120^0 \qquad \gamma_2 = 180^0$$
$$\theta_2 = 180^0 \qquad I = 2$$

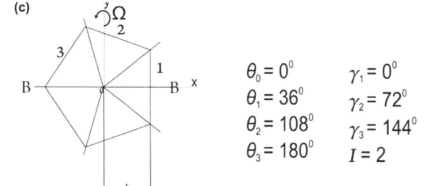

(b)

$$\theta_0 = 0^0 \qquad \gamma_1 = 45^0$$
$$\theta_1 = 90^0 \qquad \gamma_2 = 135^0$$
$$\theta_2 = 180^0 \qquad I = 2$$

(c)

$$\theta_0 = 0^0 \qquad \gamma_1 = 0^0$$
$$\theta_1 = 36^0 \qquad \gamma_2 = 72^0$$
$$\theta_2 = 108^0 \qquad \gamma_3 = 144^0$$
$$\theta_3 = 180^0 \qquad I = 2$$

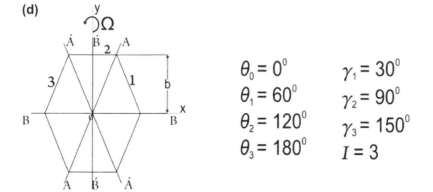

(d)

$$\theta_0 = 0^0 \qquad \gamma_1 = 30^0$$
$$\theta_1 = 60^0 \qquad \gamma_2 = 90^0$$
$$\theta_2 = 120^0 \qquad \gamma_3 = 150^0$$
$$\theta_3 = 180^0 \qquad I = 3$$

Figure 1. Rotating disc of (a) Triangle (b) Square (c) Pentagon (d) Hexagonal cross-sections.

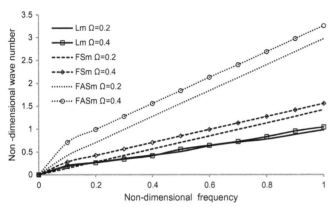

Figure 2. Variation of non-dimensional wave number vs. non-dimensional frequency of rotating disc of triangular cross-section.

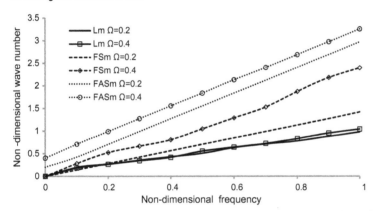

Figure 3. Variation of non-dimensional wave number vs. non-dimensional frequency of rotating disc of square cross-section.

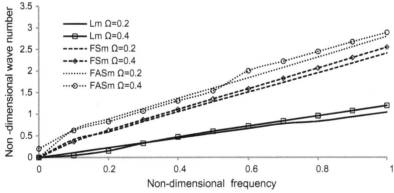

Figure 4. Variation of non-dimensional wave number vs. non-dimensional frequency of rotating disc of pentagonal cross-section.

increasing rotational speed of the disc. The flexural modes the frequency gets maximum in all the four cross-sectional discs. But, there is small oscillation and merging of curves in pentagon and hexagonal cross-section between $0 \leq \xi \leq 0.6$ in Figures 4 and 5, it shows that there is energy transportation between the modes of vibrations by the damping effect of surrounding fluid and rotational speed.

Figures 6–9 reveals that the variation of relative frequency shift of triangle, square, pentagon and hexagonal cross-sectional discs with the non-dimensional frequency for the longitudinal and flexural (symmetric and antisymmetric) modes together with different rotational speed Ω = 0.2, 0.4. From Figures 6 and 7, it is observed that, both in triangle and square cross-sections, the longitudinal and

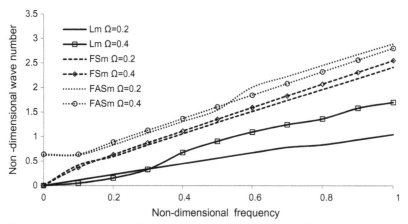

Figure 5. Variation of non-dimensional wave number vs. non-dimensional frequency of rotating disc of hexagonal cross-section.

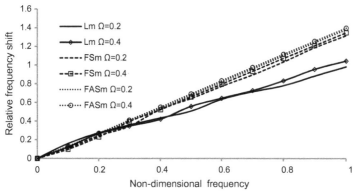

Figure 6. Variation of relative frequency shift vs. non-dimensional frequency of rotating disc of triangular cross-section.

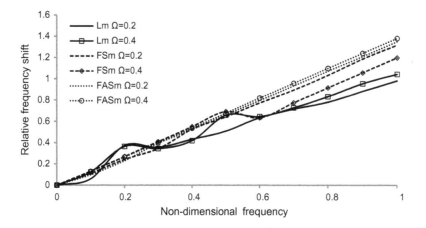

Figure 7. Variation of relative frequency shift vs. non-dimensional frequency of rotating disc of square cross-section.

flexural (symmetric and antisymmetric) modes merges for $0 \leq \xi \leq 0.2$ and begin starts increases monotonically. Also, in pentagon and hexagonal cross-sections in Figures 8 and 9, the relative frequency shift gets oscillating trend in the frequency range $0 \leq \xi \leq 0.6$. The merging of curves between the vibrational modes shows that there is energy transportation between the modes of vibrations by the effect of fluid interaction and rotation. The relative frequency shift profile is highly dispersive in trend for pentagon and hexagonal cross-section than in triangle and square cross-section.

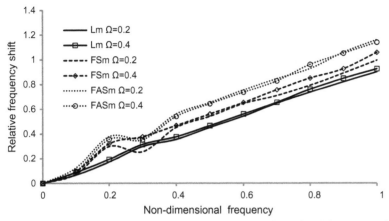

Figure 8. Variation of relative frequency shift vs. non-dimensional frequency of rotating disc of pentagonal cross-section.

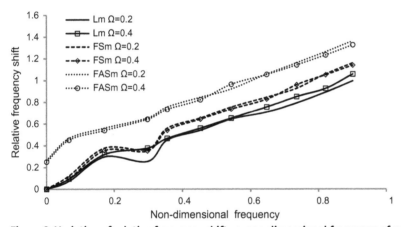

Figure 9. Variation of relative frequency shift vs. non-dimensional frequency of rotating disc of hexagonal cross-section.

10. Conclusions

In this paper, an analytical method for solving the wave propagation problem of rotating polygonal cross-sectional disc immersed in an inviscid fluid has been presented. The general frequency equation has been obtained using Fourier expansion collocation method. The frequency equations have been derived for the two cases:

(i) Polygonal cross-sectional disc with fluid and without rotation

(ii) Polygonal cross-sectional disc without fluid and rotation

and are analysed numerically for different cross-sections. Numerical calculations for non-dimensional wave number and relative frequency shift have been carried out for triangular, square, pentagonal and hexagonal cross-sectional rotating disc immersed in fluid. From the dispersion curves it is observed that the non dimensional wave number and relative frequency shift are quite higher in all the cross-sections. The effect of rotation and fluid medium on the different cross-sectional discs is also observed to be significant and more in pentagon and hexagonal cross-sections. This method is straightforward and the numerical results for any other polygonal cross-section can be obtained directly for the same frequency equation by substituting geometric values of the boundary of any cross-section analytically or numerically with satisfactory convergence.

Funding
The authors received no direct funding for this research.

Author details
R. Selvamani[1]
E-mail: selvam1729@gmail.com
[1] Department of Mathematics, Karunya University,
Coimbatore, Tamil Nadu 641114, India.

References

Abd-Alla, A. M., & Bayones, F. S. (2011). Effect of rotation in a generalized magneto thermo viscoelastic media. *Advances in Theoretical and Applied Mechanics, 4*, 15–42.

Achenbach, J. D. (1973). *Wave propagation in elastic solids*. New York, NY: Elsevier.

Bayat, M., Rahimi, M., Saleem, M., Mohazzab, A. H. M., Wudtke, I., & Talebi, H. (2014). One-dimensional analysis for magneto-thermo-mechanical response in a functionally graded annular variable-thickness rotating disk. *Applied Mathematical Modelling, 38*, 4625–4639. http://dx.doi.org/10.1016/j.apm.2014.03.008

Berliner, M. J., & Solecki, R. (1996). Wave Propagation in fluid-loaded, transversely isotropic cylinders. Part I. Analytical formulation; Part II. Numerical results. *Journal of Acoustical Society of America, 99*, 1841–1853. http://dx.doi.org/10.1121/1.415365

Bhimaraddi, A. A. (1984). A higher order theory for free vibration analysis of circular cylindrical shell. *International Journal of Solids and Structures, 20*, 623–630.

Dragomir, S. C., Sinnott, M. D., Semercigil, S. E., & Turan, Ö. F. (2014). A study of energy dissipation and critical speed of granular flow in a rotating cylinder. *Journal of Sound and Vibration, 333*, 6815–6827. http://dx.doi.org/10.1016/j.jsv.2014.07.007

Gazis, D. C. (1959). Three-dimensional investigation of the propagation of waves in hollow circular cylinders, I. Analytical foundation; II. Numerical results. *Journal of Acoustical Society of America, 31*, 568–578. http://dx.doi.org/10.1121/1.1907753

Loy, C. T., & Lam, K. Y. (1995). Vibrations of rotating thin cylindrical panels. *Applied Acoustics, 46*, 327–343. http://dx.doi.org/10.1016/0003-682X(96)81499-X

Mirsky, I. (1965). Wave propagation in transversely isotropic circular cylinders, Part I: Theory, Part II: Numerical results. *Journal of Acoustical Society of America, 37*, 1016–1026.

Nagaya, K. (1981). Dispersion of elastic waves in bars with polygonal cross section. *Journal of Acoustical Society of America, 70*, 763–770. http://dx.doi.org/10.1121/1.386914

Nagaya, K. (1983a). Vibration of thick-walled pipe or a ring of arbitrary shape in its plane. *Journal of Acoustical Society of America, 50*, 757–764.

Nagaya, K. (1983b). Vibration of a thick polygonal ring in its plane. *Journal of Acoustical Society of America, 74*, 1441–1447. http://dx.doi.org/10.1121/1.390146

Ponnusamy, P. (2011). Wave propagation in thermo-elastic plate of arbitrary cross-sections. *Multidiscipline Modeling in Materials and Structures, 7*, 1573–1605.

Ponnusamy, P., & Selvamani, R. (2012). Dispersion analysis of generalized magneto-thermoelastic waves in a transversely isotropic cylindrical panel. *Journal of Thermal Stresses, 35*, 1119–1142. http://dx.doi.org/10.1080/01495739.2012.720496

Ponnusamy, P., & Selvamani, R. (2013). Wave propagation in a transversely isotropic magneto thermo elastic cylindrical panel. *European Journal of Mechanics - A/Solids, 39*, 76–85. http://dx.doi.org/10.1016/j.euromechsol.2012.11.004

Roychoudhuri, R. S., & Mukhopadhyay, S. (2000). Effect of rotation and relaxation times on plane waves in generalized thermo visco elasticity. *IJMMS, 23*, 497–505.

Sharma, J. N., & Grover, D. (2009). Body wave propagation in rotating thermoelastic media. *Mechanics Research Communications, 36*, 715–721. http://dx.doi.org/10.1016/j.mechrescom.2009.03.005

Sinha, B. K., Plona, J., Kostek, S., & Chang, S.-K. (1992). Axisymmetric wave propagation in fluid-loaded cylindrical shells. I: Theory; II Theory versus experiment. *Journal of Acoustical Society of America, 92*, 1132–1155. http://dx.doi.org/10.1121/1.404040

Venkatesan, M., & Ponnusamy, P. (2002). Wave propagation in a solid cylinder of arbitrary cross-section immersed in fluid. *Journal of Acoustical Society of America, 112*, 936–942. http://dx.doi.org/10.1121/1.1499130

Wauer, J. (1999). Waves in rotating and conducting piezoelectric media. *Journal of Acoustical Society of America, 106*, 626–636. http://dx.doi.org/10.1121/1.427082

Yazdanpanah Moghadam, P., Tahani, M., & Naserian-Nik, A. M. (2013). Analytical solution of piezolaminated rectangular plates with arbitrary clamped/simply-supported boundary conditions under thermo-electro-mechanical loadings. *Applied Mathematical Modelling, 37*, 3228–3241. http://dx.doi.org/10.1016/j.apm.2012.07.034

Zhang, X. M. (2002). The parametric analysis of frequency of rotating laminated composite cylindrical shell using wave propagation approach. *Computer methods in applied mechanics and engineering, 191*, 2027–2043.

Appendix A

The equations for $e^1_n \sim \bar{g}^3_n$ referred in Equation 22 are as follows:

$$e^1_n = 2\left\{n(n-1)J_n(\alpha_1 ax) + (\alpha_1 ax)J_{n+1}(\alpha_1 ax)\right\}\cos 2(\theta - \gamma_l)\cos n\theta$$
$$-x^2\left\{(\alpha_1 a)^2\left(\bar{\lambda} + 2\cos^2(\theta - \gamma_l)\right)\right\}J_n(\alpha_1 ax)\cos n\theta \tag{A1}$$

$$\bar{e}^1_n = 2\left\{n(n-1)J_n(\alpha_1 ax) + (\alpha_1 ax)J_{n+1}(\alpha_1 ax)\right\}\cos 2(\theta - \gamma_l)\sin n\theta$$
$$-x^2\left\{(\alpha_1 a)^2\left(\bar{\lambda} + 2\cos^2(\theta - \gamma_l)\right)\right\}J_n(\alpha ax)\sin n\theta \tag{A2}$$

$$e^2_n = 2n\left\{(n-1)J_n(\alpha_2 ax) + (\alpha_2 ax)J_{n+1}(\alpha_2 ax)\right\}\cos 2(\theta - \gamma_l)\cos n\theta$$
$$+2\left\{\left(n(n-1)-(\alpha_2 ax)^2\right)J_n(\alpha_2 ax) + (\alpha_2 ax)J_{n+1}(\alpha_2 ax)\right\}\sin 2(\theta - \gamma_l)\sin n\theta \tag{A3}$$

$$\bar{e}^2_n = 2n\left\{(n-1)J_n(\alpha_2 ax) + (\alpha_2 ax)J_{n+1}(\alpha_2 ax)\right\}\cos 2(\theta - \gamma_l)\sin n\theta$$
$$-2\left\{\left(n(n-1)-(\alpha_2 ax)^2\right)J_n(\alpha_2 ax) + (\alpha_2 ax)J_{n+1}(\alpha_2 ax)\right\}\sin 2(\theta - \gamma_l)\cos n\theta \tag{A4}$$

$$e^3_n = \xi^2\bar{\rho}J^1_n(\alpha_3 ax)\cos n\theta \tag{A5}$$

$$\bar{e}^3_n = \xi^2\bar{\rho}J^1_n(\alpha_3 ax)\sin n\theta \tag{A6}$$

$$f^1_n = 2\left\{n(n-1)-(\alpha_1 ax)^2 J_n(\alpha_1 ax) + (\alpha_1 ax)J_{n+1}(\alpha_1 ax)\right\}\sin 2(\theta - \gamma_l)\cos n\theta$$
$$+2n\left\{(\alpha_1 ax)J_{n+1}(\alpha_1 ax) - (n-1)J_n(\alpha_1 ax)\right\}\cos 2(\theta - \gamma_l)\sin n\theta \tag{A7}$$

$$\bar{f}^1_n = 2\left\{n(n-1)-(\alpha_1 ax)^2 J_n(\alpha_1 ax) + (\alpha_1 ax)J_{n+1}(\alpha_1 ax)\right\}\sin 2(\theta - \gamma_l)\sin n\theta$$
$$-2n\left\{(\alpha_1 ax)J_{n+1}(\alpha_1 ax) - (n-1)J_n(\alpha_1 ax)\right\}\cos 2(\theta - \gamma_l)\cos n\theta \tag{A8}$$

$$f^3_n = 2n\left\{(n-1)J_n(\alpha_2 ax) - (\alpha_2 ax)J_{n+1}(\alpha_2 ax)\right\}\sin 2(\theta - \gamma_l)\cos n\theta$$
$$+2\left\{(\beta ax)J_{n+1}(\alpha_2 ax) - \left(n(n-1)-(\alpha_2 ax)^2\right)J_n(\alpha_2 ax)\right\}\cos 2(\theta - \gamma_l)\sin n\theta \tag{A9}$$

$$\bar{f}^3_n = 2n\left\{(n-1)J_n(\alpha_2 ax) - (\alpha_2 ax)J_{n+1}(\alpha_2 ax)\right\}\sin 2(\theta - \gamma_l)\sin n\theta$$
$$-2\left\{(\alpha_2 ax)J_{n+1}(\alpha_2 ax) - \left(n(n-1)-(\alpha_2 ax)^2\right)J_n(\alpha_2 ax)\right\}\cos 2(\theta - \gamma_l)\cos n\theta \tag{A10}$$

$$g^1_n = \left\{nJ_n(\alpha_1 ax) - (\alpha_1 ax)J_{n+1}(\alpha_1 ax)\right\}\cos n\theta \tag{A11}$$

$$\bar{g}^1_n = \left\{nJ_n(\alpha_1 ax) - (\alpha_1 ax)J_{n+1}(\alpha_1 ax)\right\}\sin n\theta \tag{A12}$$

$$g^2_n = nJ_n(\alpha_2 ax)\cos n\theta \tag{A13}$$

$$\bar{g}^2_n = nJ_n(\alpha_2 ax)\sin n\theta \tag{A14}$$

$$g^3_n = -\left[nJ^1_n(\alpha_3 ax) - (\alpha_3 ax)J^1_{n+1}(\alpha_3 ax)\right]\cos n\theta \tag{A15}$$

$$\bar{g}^3_n = -\left[nJ^1_n(\alpha_3 ax) - (\alpha_3 ax)J^1_{n+1}(\alpha_3 ax)\right]\sin n\theta \tag{A16}$$

10

Exergy costing analysis and performance evaluation of selected gas turbine power plants

S.O. Oyedepo[1]*, R.O. Fagbenle[2], S.S. Adefila[3] and Md.Mahbub Alam[4]

*Corresponding author: S.O. Oyedepo, Mechanical Engineering Department, Covenant University, Ota, Nigeria
E-mail: Sunday.oyedepo@ covenantuniversity.edu.ng
Reviewing editor: Duc Pham, University of Birmingham, UK

Abstract: In this study, exergy costing analysis and performance evaluation of selected gas turbine power plants in Nigeria are carried out. The results of exergy analysis confirmed that the combustion chamber is the most exergy destructive component compared to other cycle components. The exergetic efficiency of the plants was found to depend significantly on a change in gas turbine inlet temperature (GTIT). The increase in exergetic efficiency with the increase in turbine inlet temperature is limited by turbine material temperature limit. This was observed from the plant efficiency defect curve. As the turbine inlet temperature increases, the plant efficiency defect decreases to minimum value at certain GTIT (1,200 K), after which it increases with GTIT. This shows degradation in performance of gas turbine plant at high turbine inlet temperature. Exergy costing analysis shows that the combustion chamber has the greatest cost of exergy destruction compared to other components. Increasing the GTIT, both the exergy destruction and the cost of exergy destruction of this component are found to decrease. Also, from exergy costing analysis, the unit cost of electricity produced in the power plants varies from cents 1.99/kWh (N3.16/kWh) to cents 5.65/kWh (N8.98/kWh).

Subjects: Engineering Economics; Mechanical Engineering; Power & Energy

Keywords: exergy analysis; economic analysis; gas turbine; exergy cost; levelized cost; F-rule; P-rule

ABOUT THE AUTHOR

S.O. Oyedepo is presently a senior lecturer in the Department of Mechanical Engineering, Covenant University, Nigeria. His PhD research work was on *Thermodynamic Performance Analysis of Selected Gas Turbine Power Plants in Nigeria*. He has published over 45 papers in national /international journals and conferences. His major contributions in the area of energy systems and environmental noise is the publication of research and review papers in the leading journals, i.e. *Renewable and Sustainable Energy Reviews, Energy Conversion and Management, Energy Exploration & Exploitation, Environmental Monitoring Assessment and Journal of Environmental Studies*. His research interest includes but not limited to: Thermal System Design and Optimization, Energy management and Energy conversion systems, Heat Transfer Analysis, etc. Oyedepo is a registered engineer in Nigeria and member of Nigerian Society of Engineers.

PUBLIC INTEREST STATEMENT

In the last two decades, electricity-generating plants in Nigeria have been operating below their capacity with available capacity barely surpassing half the installed capacity which is short of international standards of over 95% installed capacity. Due to this low availability, other key performance indicators (capacity factor and load factor) have also been relatively low. This study therefore aims at evaluating the performance of selected gas turbine power plants in Nigeria using first and second laws of thermodynamics combined with economic analysis with a view of providing possible ways of improving the performance, thus meeting the international standards. The results of the study imply that increase in gas turbine efficiency can be achieved by improving the performance of the most inefficient component of the system. Also, the study provided a suitable methodology for relatively quick identification of the key items requiring performance improvement in a gas turbine power plant.

1. Introduction

Energy is the keystone of life and prosperity. The continued development and application of energy are essential to the sustainable advancement of society. With the exergy analysis, it is possible to evaluate the performance of energy conversion processes not only on a thermodynamics basis, but also by including economic and environmental aspects and impacts of the studied processes. This comprehensive approach of the energy resources utilization has, as one of the most important features, the identification of sustainable ways of energy resources utilization (Silvio, 2013).

The exergy analysis of thermal–mechanical conversion plants aims to characterize how the fuel exergy is used and destroyed in the energy conversion processes that take place in these plants. The exergy analysis provides means to evaluate the degradation of energy during a process, the entropy generation, and loss of opportunities to do work and thus offer space for improvement of power plant performance. Combine with economic analysis, this method allows evaluation of costs caused by irreversibility which may include the investment and operating cost of each component (Ibrahim bin, Masrul, Mohd Zamri, & Mobd, 2001).

Exergy analysis is based on the first and second laws of thermodynamics, and combines the principles of conservation of energy and non-conservation of entropy. The essence of exergy analysis is primarily for optimization. If properly done, it reveals where in the plant the largest energy wastage occurs and therefore the need for design improvements (Ofodu & Abam, 2002; Rosen, 2009). Hence, exergy is often treated as a measure of economic value (Ray, Ganguly, & Gupta, 2007).

Exergy-based cost analysis aims at determining the costs of products and irreversibilities (exergy destroyed) generated in energy conversion processes, by applying cost partition criteria which are function of the exergy content of every energy flow that takes place in the studied process. In this approach, one studies the cost formation processes by valuing the products according to its exergy content and the destroyed exergy during the energy conversion processes. This combination of exergy analysis with economic concepts is called thermoeconomic analysis when monetary costs are used and exergoeconomic analysis when exergy costs are employed (Silvio, 2013).

The needs to evaluate the cost production process in a thermal power plant can be rationally conducted if the exergy of the product of the plant (i.e. electricity generated) is taken as the value basis. This is an interesting application of exergoeconomics concepts to evaluate and allocate the cost of exergy throughout the energy conversion processes, considering costs related to exergy inputs and investment in equipment (Dincer & Rosen, 2003). Exergy is taken as a rational basis for economic cost allocation between the resources and products involved in thermal power plant processes and for the economic evaluation of their thermodynamic imperfections (Querol, Gonzalez-Regueral, & Perez-Benedito, 2013).

Exergy costing analysis is a tool used not only to evaluate the cost of inefficiencies or the costs of individual process streams (including intermediate and final products), but also to improve overall system efficiency and lower life cycle costs of a thermodynamic system (Seyyedi, Ajam, & Farahat, 2010). A complete exergoeconomic analysis consists of (1) an exergetic analysis, (2) an economic analysis, and (3) an exergoeconomic evaluation.

A number of studies on exergy and exergy costing analyses of thermal power plants have been carried out by several researchers (Ameri, Ahmadi, & Hamidi, 2009; Aras & Balli, 2008; Can, Celik, & Dagtekin, 2009; Gorji-Bandpy & Ebrahimian, 2006; Gorji-Bandpy & Goodarzian, 2011; Igbong & Fakorede, 2014; Marzouk & Hanafib, 2013; Mousafarash & Ahmadi, 2014; Mousafarash & Ameri, 2013; Kaviri, Mohd Jafar, Tholudin, & Barzegar Avval, 2011; Sahoo, 2008; Singh & Kaushik, 2014).

Most of the past studies on exergy and economic analyses of gas turbine power plant performance were based on a single gas turbine unit. In the present work, analyses are performed on 11 gas turbine units at three different stations in Nigeria.

The prime objectives of the study are:

- To evaluate the exergy performance of the selected power plants by overall exergy efficiency.
- To identify the most significant source of exergy destruction in the power plants and the location of occurrence.
- To evaluate effect of gas turbine inlet temperature (GTIT) on exergy efficiency of the selected gas turbine plants.
- To evaluate exergoeconomic performance of the selected power plants by analyzing exergetic cost parameters of each component of the power plants.
- To determine the unit cost of electricity (product) in the selected power plants using exergy costing analysis.

2. System description

Gas turbine power plants in Nigeria operate on simple gas turbine engine consisting mainly of a gas turbine coupled to a rotary type air compressor and a combustion chamber which is placed between the compressor and turbine in the fuel circuit. Auxiliaries, such as cooling fan, water pumps, etc., and the generator itself are also driven by the turbine. Other auxiliaries are starting device, lubrication system, duct system, etc. For ease of analysis, the steady state model of simple gas turbine is presented in Figure 1.

3. Methodology

Based on the idea that exergy represents the only rational basis not only for assessing the inefficiencies of thermal power system, but also for assigning costs to irreversibilities in the system, a methodological approach called exergoeconomic analysis is applied to evaluate performance of the selected gas turbine power plants in Nigeria.

3.1. Exergy costing analysis

Exergy costing analysis is an effective tool used to evaluate the cost effectiveness of thermal systems, with the intent of evaluating and enhance the system performance from both economic and exergetic (second law of thermodynamics) point of view. The analysis assists in the understanding of the cost value associated with exergy destroyed in a thermal system, and hence provides energy system's designers and operators with the information, necessary for operating, maintaining, and evaluating the performance of energy systems (Fellah, Mgherbi, & Aboghres, 2010).

In the exergy costing analysis of energy conversion system, four steps proposed by Tsatsaronis (1993) were followed in this study. The first step is exergy analysis. The second step is evaluation of non-exergy related cost (economic analysis) of each of the plant components. This step provides the monetary costs associated with investment, operation, and maintenance. The third step is the estimation of exergetic costs associated with each flow and finally, the fourth step is the exergoeconomic evaluation of each system component.

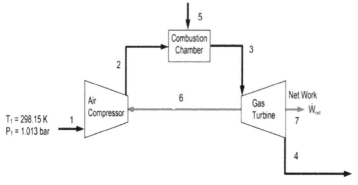

Figure 1. A schematic diagram for a simple GT cycle.

3.1.1. Exergy analysis

Exergy can be divided into four distinct components. The two important ones are the physical exergy and chemical exergy (Ahmadi, Dincer, & Rosen, 2011; Ameri et al., 2009). In this study, the other two components which are kinetic exergy and potential exergy are assumed to be negligible, as the changes in them are insignificant.

The following steps of exergy analysis itemized by Demirel (2013) are used in this study:

- Define the system boundary of processes to be analyzed.
- Define all the assumptions and the reference temperature, pressure, and composition.
- Determination of the total exergy losses.
- Determination of the thermodynamic efficiency (exergetic efficiency).
- Use exergy loss profiles to identify the regions performing poorly.
- Identification of improvements and modifications to reduce the cost of energy and operation.

Applying the first and the second laws of thermodynamics, the following exergy balance is obtained:

$$\dot{E}_x = \sum_j \left(1 - \frac{T_0}{T_j}\right)\dot{Q}_j + \dot{W}_{CV} + \sum_i m_i e_i - \sum_e m_e e_e \tag{1}$$

The subscripts i, e, j, and 0 refer to conditions at inlet and exits of control volume boundaries and reference state. $\dot{E}$, $\dot{Q}$, and $\dot{W}$ are the rates of exergy, heat, and work transfer, respectively, m is the mass flow rate. T is the absolute temperature at inlet or exit of control volume.

Equation (1) can be written as:

$$E_i^{tot} - E_e^{tot} - E_D = 0 \tag{2}$$

where $\dot{E}_i^{tot}$, $\dot{E}_e^{tot}$, and $\dot{E}_D$ are the total exergy rates at inlet and exit of control volume and rate of exergy destroyed, respectively.

Equation (2) implies that the exergy change of a system during a process is equal to the difference between the net exergy transfer through the system boundary and the exergy destroyed within the system boundaries as a result of irreversibilities.

The exergy-balance equations and the exergy destroyed during each process and for the whole gas turbine plant are written as follows (Abam & Moses, 2011):

Air compressor

$$\dot{E}^{WAC} = \left(\dot{E}_1^T - \dot{E}_2^T\right) + \left(\dot{E}_1^P - \dot{E}_2^P\right) + T_0\left(\dot{S}_1 - \dot{S}_2\right) \tag{3a}$$

$$\dot{E}_{DAC} = T_0\left(\dot{S}_2 - \dot{S}_1\right) = \dot{m}T_0\left[c_{p1-2} \ln\left(T_2/T_1\right) - R\ln\left(P_2/P_1\right)\right] \tag{3b}$$

Combustion chamber

$$\dot{E}^{CHE} + \left(\dot{E}_2^T + \dot{E}_5^T - \dot{E}_3^T\right) + \left(\dot{E}_2^P + \dot{E}_5^P - \dot{E}_3^P\right) + T_0\left(\dot{S}_3 - \dot{S}_2 + \dot{S}_5 + \frac{\dot{Q}_{CC}}{T_0}\right) = 0 \tag{4a}$$

$$\dot{E}_{DC} = T_0\left[\dot{S}_3 - \dot{S}_2 + \dot{S}_5 + \frac{\dot{Q}_{CC}}{T_0}\right] = \dot{m}T_0\left\{\left(c_{p2-3} \ln\left(T_3/T_2\right) - R\ln\left(P_3/P_2\right)\right)\right.$$

$$\left. + \left(c_{p5} \ln\left(T_5/T_0\right) - R\ln\left(P_5/P_0\right)\right) + \frac{c_{p2-3}\left(T_3 - T_2\right)}{T_{in\,CC}}\right\} \tag{4b}$$

Gas turbine

$$\dot{E}^{WGT} = \left(\dot{E}_3^T - \dot{E}_4^T\right) + \left(\dot{E}_3^P - \dot{E}_4^P\right) + T_0\left(\dot{S}_3 - \dot{S}_4\right) \tag{5a}$$

$$\dot{E}_{DGT} = \dot{m}T_0\left[c_{p3-4}\ln\left(T_4/T_3\right) - R\ln\left(P_4/P_3\right)\right] \tag{5b}$$

From Equations (3) to (5), $\dot{E}^{WAC}$ and $\dot{E}^{WGT}$ represent the exergy flow rate of the power output from the air compressor and the gas turbine, respectively. $\dot{E}_{DAC}$, $\dot{E}_{DC}$ and $\dot{E}_{DGT}$ denote the exergy destroyed in the air compressor, combustion chamber, and Gas turbine, respectively. $\dot{E}^T$ is the thermal component of the exergy stream, $\dot{E}^P$ is the mechanical component of the exergy stream, $\dot{Q}_{CC}$ represents the heat transfer rate between combustion chamber and the environment; the term $\dot{E}^{CHE}$ denotes the rate of chemical exergy flow of fuel in the combustion chamber; $\dot{S}$ is the entropy transfer rate; $T_{in\ CC}$ is the temperature of the source from which the heat is transferred to the working fluid, T_0 and P_0 are the pressure and temperature, respectively, at standard state; $\dot{m}$ is the mass flow rate of the working fluid; R is the gas constant; c_p is the specific heat at constant pressure.

For a control volume at steady state, the exergetic efficiency is

$$\varepsilon = \frac{\dot{E}_P}{\dot{E}_F} = 1 - \frac{\dot{E}_D + \dot{E}_L}{\dot{E}_F}, \tag{6}$$

where the rates at which the fuel is supplied and the product is generated are denoted by $\dot{E}_F$ and $\dot{E}^P$, respectively. $\dot{E}_D$ and $\dot{E}_L$ denote the rates of exergy destruction and exergy loss, respectively. The exergy rate of product, $\dot{E}^P$, and exergy rate of fuel, $\dot{E}_F$, for major components of gas turbine power can be determined using the equations presented by Bejan, Tsatsaronis, and Moran (1996).

The ith component efficiency defect denoted by δ_i is given by Equation (7) (Abam, Ugot, & Igbong, 2011):

$$\delta_i = \frac{\sum \Delta\dot{E}_{Di}}{\sum \Delta\dot{E}_{xin}} \tag{7}$$

where $\sum \Delta\dot{E}_{Di}$ and $\sum \Delta\dot{E}_{xin}$ are the total rate of exergy destruction in the plant and total rate of exergy flow into the plant, respectively.

The overall exergetic efficiency of the entire plant is given as:

$$\psi_i = \frac{\dot{W}_{net}}{\dot{E}_{x\ fuel}} \tag{8}$$

where $\dot{W}_{net}$ and $\dot{E}_{x\ fuel}$ are the network output of the plant and exergy of fuel (natural gas) flowing into the combustion chamber, respectively.

The amount of exergy loss rate per unit power output as important performance criteria is given as:

$$\xi = \frac{\dot{E}_{D\ Total}}{\dot{W}_{net}}, \tag{9}$$

where ξ is the exergetic performance coefficient and $\dot{E}_{D\ Total}$ denotes the total exergy destroyed in the entire plant.

Exergy destruction rate and efficiency equations for the gas turbine power plant components and for the whole cycle are summarized in Table 1.

Table 1. The exergy destruction rate and exergy efficiency equations for gas turbine		
Component	**Exergy destruction rate**	**Exergy efficiency**
Compressor	$\dot{E}_{DC} = \dot{E}_{in} - \dot{E}_{out} + \dot{W}_C$	$\varepsilon = \frac{\dot{E}_{out} - \dot{E}_{in}}{\dot{W}}$
Combustion chamber	$\dot{E}_{DCC} = \dot{E}_{in} - \dot{E}_{out} + \dot{E}_{fuel}$	$\varepsilon = \frac{\dot{E}_{out}}{\dot{E}_{in} - \dot{E}_{fuel}}$
Gas turbine	$\dot{E}_{DT} = \dot{E}_{in} - \dot{E}_{out} - (\dot{W}_{net} + \dot{W}_C)$	$\varepsilon = \frac{\dot{W}_{net} + \dot{W}_C}{\dot{E}_{in} - \dot{E}_{out}}$

Total exergy destruction rate $\dot{E}_{D\,Total} = \sum \dot{E}_D = \dot{E}_{DC} + \dot{E}_{DCC} + \dot{E}_{DT}$

3.1.2. Estimation of non-exergy related cost

The economics of gas turbine assess the non-exergy related cost; which is the cost of the various components of the system (Igbong & Fakorede, 2014). This cost comprises the cost associated with the investment, operation, maintenance, and fuel costs of gas turbine power plant (Ahmadi & Dincer, 2011; Siahaya, 2009). These monetary values are used in the cost balances to determine cost flow rates (Bejan et al., 1996).

The annualized (levelized) cost method of Moran (1982) is used to estimate the capital cost of system component in this work.

The amortization cost for a particular component may be written as (Kim, Oh, Kwon, & Kwak, 1998):

$$PW = PEC - (SV)\ PWF\ (i, n), \tag{10}$$

where the salvage value (SV) at the end of the nth year is taken as 10% of the initial investment for component (or purchase equipment cost, PEC). The present worth (PW) of the component may be converted to the annualized cost by using the capital recovery factor CRF (i, n) (Gorji-Bandpy & Goodarzian, 2011; Kim et al., 1998), i.e.

$$\dot{C}\ (\$/year) = PW \times CRF\ (i, n), \tag{11a}$$

$$CRF\ (i, n) = i/1 - (1 + i)^{-n}, \tag{11b}$$

$$PWF = (1 + i)^{-n}, \tag{11c}$$

where i is the interest rate and it is taken to be 17% (Gorji-Bandpy & Goodarzian, 2011), n is the total operating period of the plant in years and was obtained from the selected plants. PEC is the purchased equipment cost and $\dot{C}$ is the annualized cost of the component.

Equations for calculating the PEC for the components of the gas turbine power plant are as follows (Barzegar Avval, Ahmadi, Ghaffarizadeh, & Saidi, 2011; Bejan et al., 1996; Gorji-Bandpy, Goodarzian, & Biglari, 2010):

Air compressor

$$PEC_{ac} = \left[\frac{71.1\dot{m}_a}{0.9 - \eta_{sc}}\right]\left[\frac{P_2}{P_1}\right]\ln\left[\frac{P_2}{P_1}\right] \tag{12}$$

Combustion chamber

$$PEC_{cc} = \left[\frac{46.08\dot{m}_a}{0.995 - P_3/P_2}\right][1 + \exp(0.018T_3 - 26.4)] \tag{13}$$

Gas turbine

$$PEC_{gt} = \left[\frac{479.34\dot{m}_g}{0.92 - \eta_{st}}\right] \ln\left[\frac{P_3}{P_4}\right][1 + \exp(0.036T_3 - 54.4)] \tag{14}$$

Dividing the levelized cost by annual operating hours, N, we obtain capital cost rate for the kth component of the plant (Kwon, Kwak, & Oh, 2001):

$$\dot{Z}_k = \frac{\phi_k \dot{C}_k}{3600 \times N} \tag{15}$$

From Equations (12) to (15), PEC_{ac}, PEC_{cc}, and PEC_{gt} represent the purchasing equipment cost for air compressor, combustion chamber, and gas turbine component, respectively; $\dot{m}_a$ and $\dot{m}_g$ denote air mass flow rate and mass flow rate of gas product in the plant, respectively; η_{sc} and η_{st} are the isentropic efficiencies of compressor and gas turbine, respectively; P_1, P_2, P_3, and P_4 are compressor inlet pressure, compressor outlet pressure, combustor inlet pressure, and gas turbine inlet pressure, respectively; T_3 and T_4 represent the combustor inlet temperature and combustor outlet temperature, respectively; ϕ_k is maintenance factor.

The maintenance cost is taken into consideration through the factor ϕ_k = 1.06 for each plant component (Gorji-Bandpy & Goodarzian, 2011; Gorji-Bandpy et al., 2010).

3.1.3. The auxiliary equation rules
According to Lazzaretto and Tsatsaronis (1999), the following two rules for formulating the auxiliary equations are valid, when finding the specific costs of exergy associated with flows is desired. These are:

F-principle

The total cost associated with the removal of exergy must be equal to the cost at which the removed exergy was supplied to the same stream in upstream components.

P-principle

Each exergy unit is supplied to any stream associated with the product at the same average cost.

3.1.4. Estimation of gas turbine exergetic costs
Exergy costing is usually applied at the plant component (Tsatsaronis & Winhold, 1984). In order to perform exergy costing calculations, gas turbine components (Figure 1) must be combined into suitable control volumes on which exergetic cost balance equation was then applied on an individual basis. The component in each control volume (CV) with their input and output streams are given as follows:

CV 1: Air compressor (AC)—Input streams: 1, 6
Output stream: 2
CV 2: Combustion chamber (CC)—Input streams: 2, 5
Output stream: 3
CV 3: Gas turbine (GT))—Input stream: 3
Output streams: 4, 6, 7

For a component that receives heat transfer and generates power, cost balance equation may be written as follows (Ameri et al., 2009; Barzegar Avval et al., 2011; Gorji-Bandpy & Goodarzian, 2011; Gorji-Bandpy et al., 2010):

$$\sum(c_e\dot{E}_e)_k + c_{w,k}\dot{W}_k = c_{q,k}\dot{E}_{q,k} + \sum(c_i\dot{E}_i)_k + \dot{Z}_k \tag{16}$$

$$\dot{C}_j = c_j \dot{E}_j \tag{17}$$

The cost–balance equations for all the components of the system construct a set of nonlinear algebraic equations, which was solved for $\dot{C}_j$ and c_j.

The formulations of cost balance for each component and the required auxiliary equations are as follows:

Air compressor

$$\dot{C}_2 = \dot{C}_1 + \dot{C}_6 + \dot{Z}_{ac} \tag{18}$$

where subscript 6 denotes the power input to the compressor.

Combustion chamber

$$\dot{C}_3 = \dot{C}_2 + \dot{C}_5 + \dot{Z}_{cc} \tag{19}$$

Gas turbine

$$\dot{C}_4 + \dot{C}_6 + \dot{C}_7 = \dot{C}_3 + \dot{Z}_{gt} \tag{20}$$

The auxilia.ry equation for gas turbine is given as:

$$\frac{\dot{C}_3}{\dot{E}_3} = \frac{\dot{C}_4}{\dot{E}_4} \tag{21}$$

Additional auxiliary equation is formulated assuming the same unit cost of exergy for the net power exported from the system and power input to the compressor:

$$\frac{\dot{C}_6}{\dot{W}_{AC}} = \frac{\dot{C}_7}{\dot{W}_n} \tag{22}$$

From Equations (16) to (22), $\dot{Z}_k$ is capital cost rate of unit k; c_i and c_e represent cost per exergy unit at inlet and exit of component k; $\dot{E}_i$ and $\dot{E}_e$ are exergy flow rates at inlet and exit of component k, respectively; $\dot{C}_j$ is monetary flow rate of material stream j; $\dot{Z}_{ac}$, $\dot{Z}_{cc}$, and $\dot{Z}_{gt}$ denote capital cost rates of air compressor, combustion chamber, and gas turbine, respectively; $\dot{W}_n$ and $\dot{W}_{AC}$ represent network output and compressor work input, respectively.

The cost rate associated with fuel (methane) is obtained from (Valero et al., 1994):

$$\dot{C}_f = c_f \dot{m}_f \times LHV \tag{23}$$

where the fuel cost per energy unit (on an LHV basis) is $c_f = 0.004$ \$/MJ (Valero et al., 1994), $\dot{m}_f$ is the mass flow rate of fuel and LHV is the lower heating value of fuel.

A zero unit cost is assumed for air entering the air compressor, i.e.

$$\dot{C}_1 = 0 \tag{24}$$

In order to estimate the cost of exergy destruction in each component of the plant, the cost–balance equations were solved for each component. In application of the cost–balance equation (Equation 17), there is usually more than one inlet and outlet streams for some components. In this case, the numbers of unknown cost parameters are higher than the number of cost -balance equations for that component. Auxiliary exergoeconomic equations (Equations 22 and 23) are developed to solve this problem. Implementing Equation (17) for each component together with the auxiliary equations forms a system of linear equations as follows (Ahmadi, Barzegar, Ghaffarizadeh, & Saidi, 2010; Ahmadi, Dincer, et al., 2011; Ameri, Ahmadi, & Khanmohammadi, 2008):

$$[\dot{E}_k] \times [c_k] = [\dot{Z}_k] \tag{25}$$

where $[\dot{E}_k]$, $[c_k]$, and $[\dot{Z}_k]$ are the matrix of exergy rate which were obtained in exergy analysis, exergetic cost vector (to be evaluated) and the vector of $\dot{Z}_k$ factors (obtained in economic analysis), respectively.

The above set of equations was solved using MATLAB to obtain the cost rate of each line in Figure 1.

3.1.5. Exergoeconomic variables for gas turbine components evaluation

In exergoeconomic evaluation of thermal systems, certain quantities play an important role. These are the average cost of fuel ($c_{F,k}$), average unit cost of product ($c_{P,k}$), the cost rate of exergy destruction ($\dot{C}_{D,k}$), relative cost difference r_k and exergoeconomic factor f_k (Ahmadi, Ameri, & Hamidi, 2009; Gorji-Bandpy et al., 2010).

Then the average costs per unit of fuel exergy ($c_{F,k}$) and product exergy ($c_{P,k}$) are calculated from (Fellah et al., 2010):

$$c_{F,k} = \frac{\dot{C}_{F,k}}{\dot{E}_{F,k}} \tag{26}$$

$$c_{P,k} = \frac{\dot{C}_{P,k}}{\dot{E}_{P,k}} \tag{27}$$

The cost rate associated with exergy destruction is estimated as:

$$\dot{C}_{D,k} = c_{F,k}\dot{E}_{D,k} \tag{28}$$

Relative cost difference r_k is given as (Moran & Tsatsaronis, 2000):

$$r_k = \frac{c_{P,k} - c_{F,k}}{c_{F,k}} = \frac{1 - \varepsilon_k}{\varepsilon_k} + \frac{\dot{Z}_k}{c_{F,k}\dot{E}_{p,k}} \tag{29}$$

One indicator of exergoeconomic performance is the exergoeconomic factor, f_k. The exergoeconomic factor is defined as (Fellah et al., 2010; Gorji-Bandpy & Goodarzian, 2011):

$$f_k = \frac{\dot{Z}_k}{\dot{Z}_k + \dot{C}_{D,k}} \tag{30}$$

4. Results and discussion

The average operating data for the selected gas turbine power plants for the period of six years (2005–2010) are presented in Table 2.

4.1. Results of exergy analysis

The exergy flow rates at the inlet and outlet of each component of the plants were evaluated based on the values of measured properties such as pressure, temperature, and mass flow rates at various states. These quantities were used as input data to the computer program (MATLAB) written to perform the simulation of the performance of the components of the gas turbine power plant and the overall plant.

Table 2. Average operating data for selected gas turbine power plants

Plant/ average operating data	AES station			Afam station				Delta station			
	PB204 (AES1)	PB209 (AES2)	PB210 (AES3)	GT17 (AF1)	GT18 (AF2)	GT19 (AF3)	GT20 (AF4)	GT9 (DEL1)	GT10 (DEL2)	GT18 (DEL3)	GT20 (DEL4)
Ambient temperature, T_1 (K)	303.63	302 31	305.28	300.34	301.48	300.38	300.9	300.55	301.41	301.15	301.79
Compressor outlet temperature, T_2 (K)	622.31	627.48	636.28	593.73	595.82	610.90	618.32	613.73	619.07	634.32	630.32
Turbine inlet temperature, T_3 (K)	1,218.62	1,256.86	1 222 45.	1.133,4	1,192.82	1,200.15	1,215.65	1,226.15	1,224.73	1,233.57	1,234.73
Turbine outlet temperature, T_4 (K)	750.00	755 00	827 05.	712.73	723.75	770.07	807.32	710.56	707.48	730.15	705.07
Temperature of exhaust gas, T_{exh} (K)	715.40	750.52	746 48.	731.45	664.65	707.23	741.48	622.65	649.90	635.65	636.73
Compressor inlet pressure, P_1 (bar)	1.013	1 013	1.013	1.013	1.013	1.013	1.013	1.013	1.013	1.013	1.013
Compressor outlet pressure, P_2 (bar)	9.8	9.86	9.60	9.50	9.80	9.60	9.60	11.05	10.98	10.82	10.84
Pressure ratio	9.00	9.14	9 48.	9.38	9.67	9.48	9.48	10.91	10.84	10.68	10.70
Mass flow rate of fuel (kg/s)	2.58	2 54	2.81	6.50	6.40	8.10	8.40	3.08	3.10	8.15	8.13
Inlet mass flow rate of air (kg/s)	122.16	122 20	121 93.	359.00	359.00	470	470	140	140	375	375
Power output (MW)	29.89	29.37	31.52	49.90	58.00	132	135.4	19.42	20.8	92.8	93.42
LHV of fuel (kJ/kg)	47,541.57	47,541.57	47 541 57.	48.948,3	48,948.3	48,948.3	48,948.3	46,778	46,778	46,778	46,778

Table 3 presents results of the net exergy flow rates crossing the boundary of each component of the plants, exergy destruction, exergy defect, exergetic performance coefficient, and exergy efficiency of each component of the plants. The two most important performance criteria, exergy efficiency and exergetic performance coefficient (ξ), vary from 18.22 to 32.84% and 1.45 to 2.44%, respectively, for the considered plants. Since the condition of good performance is derived from a higher overall exergetic efficiency but a lower exergetic performance coefficient for any thermal system; hence, it can be inferred that AF2, AF1, and DEL4 gas turbine plants have good performance. The total exergy destruction rates vary from 59.42 to 234.49 MW, AF4 has the highest value and DEL2 has the least value. The total efficiency defects and overall exergetic efficiency vary from 38.64 to 69.32% and 15.66 to 30.72%, respectively. The efficiency defects are higher for AF4 (69.32%) and AF3 (62.94%) than other units.

The exergy analysis results also show that the highest percentage exergy destruction occurs in the combustion chamber (CC) and followed by the air compressor in some plants in the range of 82.61–91.29% and 4.10–8.16%, respectively. Hence, the combustion chamber is the major source of thermodynamic inefficiency in the plants considered due to the irreversibility associated with combustion and the large temperature difference between the air entering the combustion chamber and the flame temperature. These immense losses basically mean that a large amount of energy present in the fuel, with great capacity to generate useful work, is being wasted. The variations in performance of the plants may be ascribed to poor maintenance procedures, faulty components, and discrepancies in operating data.

Table 3. Result of exergy analysis

Exergy performance indicator	PB204 (AES1)	PB209 (AES2)	PB210 (AES3)	GT17 (AF1)	GT18 (AF2)	GT19 (AF3)	GT20 (AF4)	GT9 (DEL1)	GT10 (DEL2)	GT18 (DEL3)	GT20 (DEL4)
Installed rated power (MW)	33.5	33.5	33.5	75.0	75.0	138.0	138.0	25.0	25.0	100.0	100.0
Fuel exergy flow rate (MW)	220.53	235.23	237.68	327.96	363.28	459.15	449.06	274.85	276.78	441.20	440.24
Exergy destruction rate of AC (MW)	4.69	4.98	5.64	8.62	8.09	13.14	14.80	3.14	3.63	13.36	12.48
Exergy destruction of CC (MW)	56.55	56.58	55.35	139.42	159.84	176.78	180.83	62.52	61.76	171.84	173.33
Exergy destruction rate of turbine (MW)	0.29	0.52	0.23	5.99	1.47	9.39	14.50	0.39	0.14	0.70	1.80
Total exergy destruction rate (MW)	61.54	62.09	61.23	154.02	169.40	199.31	210.13	66.04	65.53	185.91	187.61
Exergy destruction of AC (%)	7.62	8.03	9.21	5.59	4.78	6.60	7.04	4.75	5.54	7.19	6.65
Exergy destruction of CC (%)	91.90	91.13	90.39	90.51	94.36	88.70	86.05	94.67	94.25	92.43	92.39
Exergy destruction rate of turbine (%)	0.48	0.84	0.41	3.89	0.87	4.71	6.90	0.58	0.21	0.38	0.96
Efficiency defect of AC (%)	14.01	14.83	16.83	9.15	8.43	12.46	14.03	7.79	9.05	12.52	11.69
Efficiency defect of CC (%)	66.11	66.26	64.63	58.31	58.05	58.35	56.20	73.45	72.43	56.11	56.97
Efficiency defect of turbine (%)	0.38	0.68	0.32	3.05	0.68	4.08	6.50	0.42	0.15	0.29	0.74
Total efficiency defect (%)	80.50	81.77	81.78	70.51	67.16	74.89	76.73	81.66	81.63	68.92	69.40
Exergy efficiency of AC (%)	85.99	85.17	83.17	90.85	91.57	87.54	85.97	95.21	90.95	87.48	88.31
Exergy efficiency of CC (%)	74.36	75.95	76.71	57.49	56.00	61.50	59.73	77.25	77.69	61.05	60.63
Exergy efficiency of turbine (%)	99.62	99.32	99.67	96.86	99.32	95.75	93.09	99.57	99.85	99.71	99.25
Overall exergetic efficiency (%)	19.50	18.23	18.22	29.49	32.84	25.11	23.27	18.34	18.37	31.08	30.60
Exergetic performance coefficient (ξ)	1.43	1.45	1.46	1.59	1.32	1.73	2.01	1.47	1.59	1.36	1.39
Exergy improvement potential											
Air compressor (MW)	3.94	4.14	4.71	7.83	6.88	8.59	14.95	2.89	3.30	11.69	11.02
Combustion chamber (MW)	38.48	40.33	43.82	69.49	68.52	85.81	88.86	30.47	30.21	78.48	78.67
Turbine (MW)	0.015	0.023	0.13	0.19	0.22	1.05	2.50	0.15	0.17	1.00	0.38
Entire plant (MW)	119.44	147 15	159 28	98 30	89 99	124 18	59 99	54 04	56 46	159 88	143 28

To illustrate the effect of operating parameters on the second law efficiency of the components of the gas turbine, the AES1 (PB204) plant is considered as a typical case. The simulation of the performance of plant and components was done by varying the air inlet temperature from 290 to 320 K; and the turbine inlet temperature from 1,000 to 1,400 K, respectively. Figure 2 compares the second law efficiencies of the air compressor, combustion chamber, gas turbine, and the overall plant when the ambient temperature increases. The exergy efficiency of the turbine component and the overall exergetic efficiency of plant decreased with increased ambient temperature, whereas the exergy efficiencies of the compressor and turbine increased with increased ambient temperature. The

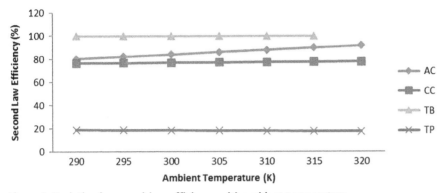

Figure 2. **Variation in second-law efficiency with ambient temperature.**

Notes: AC: Second law efficiency of compressor, CC: Second law efficiency of combustion chamber, TB: Second law efficiency of turbine, TP: Second law efficiency of entire plant.

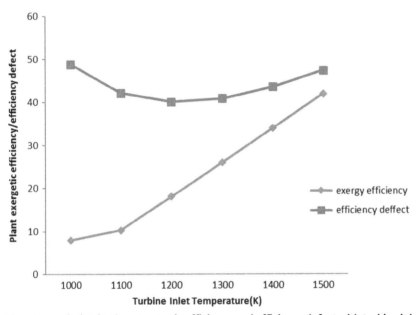

Figure 3. **Variation in plant exergetic efficiency and efficiency defect with turbine inlet temperature.**

overall exergetic efficiency decreased from 18.53 to 17.26% for ambient temperature range of 290–320 K. It was found that a 5 K rise in ambient temperature resulted in a 1.03% decrease in the overall exergetic efficiency of the plant. The reason for the low overall exergetic efficiency is due to large exergy destruction in the combustion chamber (Kotas, 1995).

The exergetic efficiency (or second law efficiency) of the plant was also found to depend significantly on a change in turbine inlet temperature. Figure 3 shows that the second law efficiency of the plant increases steadily as the turbine inlet temperature increases. The increase in exergetic efficiency with the increase in turbine inlet temperature is limited by turbine material temperature limit. This can be seen from the plant efficiency defect curve. As the turbine inlet temperature increases, the plant efficiency defect decreases to minimum value at certain TIT (1,200 K), after which it increases with TIT. This shows degradation in performance of gas turbine plant at high turbine inlet temperature.

4.2. Results of exergy costing analysis

Solving the linear system of Equations (14–20), the cost rates of the unknown streams of the system are obtained. Tables 4–6 show the results of levelized cost rates and average costs per unit of exergy at various state points in the AES Barges gas turbine system, Afam gas turbine system, and Delta gas

Table 4. Levelized cost rates and average costs per unit of exergy at various state points in the AES Barges gas turbine system

State points	AES1 (PB204)			AES2 (PB209)			AES3 (PB210)		
	$\dot{C}$ ($/h)	C ($/GJ)	c ($/kWh)	$\dot{C}$ ($/h)	C ($/GJ)	c ($/kWh)	$\dot{C}$ ($/h)	C ($/GJ)	c ($/kWh)
1	0.00	0.00	0.00	0.00	0.00	0.00	0.00	0.00	0.00
2	1,536.70	11.67	0.042011	2,001.24	15.03	0.054108	2,098.72	15.67	0.056422
3	4,001.51	10.69	0.038468	5,167.1	13.65	0.049153	5,517.8	13.65	0.049128
4	1,117.72	10.69	0.038468	1,459.99	13.65	0.049153	1,825.40	13.65	0.049128
5	3,080.69	3.08	0.011090	3,491.45	3.49	0.012569	3,628.37	3.63	0.013062
6	1,536.58	10.73	0.038617	2,001.10	13.75	0.049489	2,098.59	14.07	0.050655
7	1,347.31	10.73	0.038617	1,706.18	13.75	0.049489	1,593.96	14.07	0.050655

Table 5. Levelized cost rates and average costs per unit of exergy at various state points in the Afam gas turbine system

State points	AF1 (GT17)			AF2 (GT18)			AF3 (GT19)			AF4 (GT20)		
	$\dot{C}$ ($/h)	C ($/GJ)	c ($/kWh)	$\dot{C}$ ($/h)	C ($/GJ)	c ($/kWh)	$\dot{C}$ ($/h)	C ($/GJ)	c ($/kWh)	$\dot{C}$ ($/h)	C ($/GJ)	c ($/kWh)
1	0.00	0.00	0.00	0.00	0.00	0.00	0.00	0.00	0.00	0.00	0.0	0.00
2	3,807.70	0.144	0.037594	3,504.63	9.54	0.034339	5,562.00	11.73	0.042246	7,475.78	14.95	0.053813
3	9,254.61	9.62	0.034631	8,784.92	8.86	0.031909	14,019.99	10.59	0.038114	16,571.89	12.42	0.044695
4	2,436.02	9.62	0.034631	2,346.86	8.86	0.031909	4,311.84	10.59	0.038114	5,683.35	12.42	0.044695
5	4,581.56	4.58	0.016494	4,511.08	4.51	0.016240	5,709.33	5.71	0.020554	5,920.79	5.92	0.021315
6	3,807.29	9.92	0.035711	3,504.11	9.16	0.032977	5,561.33	11.29	0.040645	7,475.00	13.73	0.049442
7	3,011.59	9.92	0.035711	2,934.32	9.16	0.032977	4,147.30	11.29	0.040645	3,414.10	13.73	0.049442

Table 6. Levelized cost rates and average costs per unit of exergy at various state points in the Delta gas turbine system

State points	DEL1 (GT9)			DEL2 (GT10)			DEL3 (GT18)			DEL4 (GT20)		
	$\dot{C}$ ($/h)	C ($/GJ)	c ($/kWh)	$\dot{C}$ ($/h)	C ($/GJ)	c ($/kWh)	$\dot{C}$ ($/h)	C ($/GJ)	c ($/kWh)	$\dot{C}$ ($/h)	C ($/GJ)	c ($/kWh)
1	0.00	0.00	0.00	0.00	0.00	0.00	0.00	0.00	0.00	0.00	0.00	0.00
2	835.48	5.43	0.019552	898.54	5.80	0.020875	7,151.43	16.86	0.060689	6,588.44	15.63	0.056263
3	1,953.22	5.01	0.018036	2,030.82	5.28	0.019014	14,976.74	14.63	0.052653	14,378.32	14.02	0.050462
4	494.52	5.01	0.018036	514.49	5.28	0.019014	4,145.25	14.63	0.052653	3,621.36	14.02	0.050462
5	2,074.70	2.07	0.007469	2,088.17	2.09	0.007517	5,489.87	5.49	0.019764	5,476.39	5.48	0.019715
6	835.31	5.23	0.018818	898.32	5.53	0.019900	7,150.92	15.68	0.056460	6,587.93	14.62	0.052639
7	623.50	5.23	0.018818	618.15	5.53	0.019900	3,680.89	15.68	0.056460	4,169.36	14.62	0.052639

turbine system, respectively. For these systems, the unit cost of electricity computed from the exergy costing method for the selected plants varies from $5.23/GJ (cents 1.88/kWh) (N2.99/kWh) to $15.68/GJ (cents 5.65/kWh) (N8.98/kWh) (see Tables 4–6).

The exergoeconomic parameters for each of the components of the plants considered in this study for their actual operating conditions are summarized in Tables 7–9. The parameters include average costs per unit of fuel exergy C_F and product exergy C_p, rate of exergy destruction $\dot{E}_D$, cost rate of exergy destruction $\dot{C}_D$, investment, and O &M costs rate $\dot{Z}$ and exergoeconomic factor f. In analytical terms, the components with the highest value of $\dot{Z}_k + \dot{C}_{Dk}$ are considered the most significant components from an exergoeconomic perspective (Gorji-Bandpy et al., 2010). This provides a means of determining the level of priority a component should be given with respect to the improving of the system.

Table 7. Exergoeconomic parameters of the gas turbine components for AES barges station

Exergoeconomic parameters	AES1 (PB204)			AES 2 (PB20 9)			AES 3(PB210)		
	AC	CC	GT	AC	CC	GT	AC	CC	GT
C_F ($/GJ)	10.73	4.96	10.69	13.75	5.28	13.65	1 4.07	5.32	13.65
C_p ($/GJ)	11.67	10.69	10.72	15.03	13.67	13.72	1 5.67	13.65	13.93
$\dot{E}_D$ (MW)	3.21	53.83	0.29	3.45	53.82	0.51	4.23	59.92	2.26
$\dot{C}_D$ ($/h)	124.01	2,761.63	11.06	170.72	3,493.79	25.04	2 14.41	3,575.46	111.18
$\dot{Z}$ ($/h)	126.22	12.56	87.19	145.29	14.35	100.24	1 23.38	12.93	87.54
$\dot{C}_D + \dot{Z}$ ($/h)	250.23	2,774.19	98.25	316.01	3,508.13	125.27	3 37.79	3,588.39	198.72
f (%)	50.44	0.45	88.74	45.98	0.41	80.01	3 6.53	0.36	44.05

Table 8. Exergoeconomic parameters of the gas turbine components for Afam gas turbine station

Exergoeconomic parameters	AF1 (GT17)			AF2 (GT18)			AF3 (GT19)			AF4 (GT20)		
	AC	CC	GT	AC	CC	GT	AC	CC	GT	AC	CC	GT
C_F ($/GJ)	9.92	5.40	9.62	9.16	5.21	8.86	11.29	5.76	10.59	13.73	6.58	12.42
C_p ($/GJ)	10.44	9.62	9.84	9.54	8.86	9.08	11.73	10.59	11.06	14.95	12.42	13.25
$\dot{E}_D$ (MW)	5.33	136.13	5.95	4.20	143.18	6.52	5.17	197.12	15.85	12.27	188.59	23.38
$\dot{C}_D$ ($/h)	190.33	3,193.95	205.97	138.43	2,849.94	08.21	210.09	3,638.03	604.19	606.50	4,615.55	1,044.89
$\dot{Z}$ ($/h)	403.02	41.58	291.66	520.31	51.38	63.97	674.10	68.66	482.37	776.95	79.18	556.31
$\dot{C}_D + \dot{Z}$ ($/h)	593.35	3,235.53	497.63	658.74	2,901.32	75.18	884.19	3,706.70	1,086.57	1,383.44	4,694.72	1,601.20
f (%)	67.92	1.29	58.61	78.99	1.77	63.68	76.24	1.85	44.39	56.16	1.69	34.74

Table 9. Exergoeconomic parameters of the gas turbine components for Delta gas turbine station

Exergoeconomic parameters	DEL1 (GT9)			DEL2 (GT10)			DEL3 (GT18)			DEL4 (GT20)		
	AC	CC	GT	AC	CC	GT	AC	CC	GT	AC	CC	GT
C_F ($/GJ)	5.23	4.20	5.01	5.53	4.28	5.28	15.68	6.83	14.63	14.62	6.54	14.02
C_p ($/GJ)	5.43	5.010	5.17	5.80	5.28	5.46	16.86	14.63	15.38	15.63	14.02	14.46
$\dot{E}_D$ (MW)	1.66	52.62	3.35	2.10	50.71	3.55	8.82	130.36	13.86	8.05	131.82	8.81
$\dot{C}_D$ ($/h)	31.18	1,272.00	60.51	41.74	1,339.48	67.43	497.89	5,642.82	729.92	423.84	5,353.55	444.60
$\dot{Z}$ ($/h)	168.06	13.97	104.96	216.99	18.19	136.40	508.98	43.57	324.53	510.68	43.61	325.00
$\dot{C}_D + \dot{Z}$ ($/h)	199.24	1,285.98	165.47	258.72	1,357.67	203.83	1,006.87	5,686.39	1,054.45	934.52	5,397.15	769.60
f (%)	84.35	1.09	63.43	83.87	1.34	66.92	50.55	0.77	30.78	54.65	0.81	42.23

For all the plants considered, the combustion chamber and air compressor have the highest value of the sum $\dot{Z}_k + \dot{C}_{Dk}$ and are, therefore, the most important components from the exergoeconomic viewpoint. The low value of exergoeconomic factor, f, associated with the combustion chamber suggests that the cost rate of exergy destruction is the dominate factor influencing the component. Hence, it is implied that the component efficiency is improved by increasing the capital investment. This can be achieved by increasing GTIT. The maximum GTIT is limited by the metallurgical considerations (Altayib, 2011; Gorji-Bandpy & Goodarzian, 2011).

Considering air compressor which has the second highest value of the sum $\dot{Z}_k + \dot{C}_{Dk}$ (except for units AF3, AF4 and DEL3), the relatively large value of the factor f suggests that the capital investment and O and M costs dominate. According to Equation (15) of the cost model, the capital investment costs of the air compressor depend on pressure ratio (P_2/P_1) and compressor isentropic efficiency η_{sc}. Reduction of the $\dot{Z}$ value associated with the air compressor may be achieved by

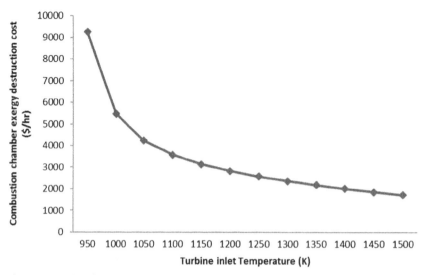

Figure 4. Combustion chamber exergy destruction cost and TIT.

reducing the pressure ratio (P_2/P_1) and /or the isentropic efficiency η_{sc} (Bejan et al., 1996). Moreover, Tables 7–9 show that the exergy destruction and investment cost are almost equal for air compressor when compared with other components. This implies that the systems performance may be improved by increasing the investment cost of this component.

The results of the exergy costing analysis of the plants investigated show that the combustion chamber (CC) exhibits the greatest exergy destruction cost. The next highest source of exergy destruction cost is the air compressor. In comparing the results of exergy and exergy costing analyses, similar trends are revealed. Increasing GTIT effectively decreases the cost associated with exergy destruction. Further comparisons between related results are consistent with those reported by Ahmadi, Rosen, and Dincer (2011), and confirm that the most significant parameter in the plant is GTIT. The finding establishes the concept that the exergy loss in the combustion chamber is associated with the large temperature difference between the flame and the working fluid. Reducing this temperature difference reduces the exergy loss. Furthermore, cooling compressor inlet air allows the compression of more air per cycle, effectively increasing the gas turbine capacity.

To illustrate the effect of GTIT on the exergy destruction cost of combustion chamber of the selected plants, AES1 (PB204) plant is considered as sample. The simulation was done by varying the GTIT from 950 to 1,500 K. Figure 4 shows the effect of variation in GTIT on combustion chamber exergy destruction cost. This figure shows that, like the exergy analysis results, the cost of exergy destruction for the combustion chamber decreases with an increase in the GTIT. This is due to the fact that the cost of exergy destruction is proportional to the exergy destruction. Hence, an increase in the GTIT can decrease the cost of exergy destruction. Furthermore, from Figure 3, an increase in the TIT of about 200 K can lead to a reduction of about 29% in the cost of exergy destruction. Therefore, TIT is the best option to improve cycle losses.

5. Conclusions and recommendations

In the present study, exergy costing analysis and performance assessment from thermodynamics point of view were performed for 11 selected gas turbine power plants in Nigeria.

The results from the exergy analysis show that the combustion chamber is the most significant exergy destructor in the selected power plants, which is due to the chemical reaction and the large temperature difference between the burners and working fluid. Moreover, the results show that an increase in GTIT leads to an increase in gas turbine exergy efficiency due to a rise in the output power of the turbine and a decrease in the combustion chamber losses.

The results from the exergy costing analysis, in common with those from the exergy analysis, show that the combustion chamber has the greatest cost of exergy destruction compared to other components. In addition, by increasing GTIT, the gas turbine cost of exergy destruction can be decreased. The finding solidifies the concept that the exergy loss in the combustion chamber is associated with the large temperature difference between the flame and the working fluid. Reducing this temperature difference reduces the exergy loss. Furthermore, cooling compressor inlet air allows the compression of more air per cycle, effectively increasing the gas turbine capacity. The results of this study revealed that an increase in the GTIT of about 200 K can lead to a reduction of about 29% in the cost of exergy destruction. Therefore, GTIT is the best option to improve cycle losses. From exergy costing analysis, the unit cost of electricity produced in the selected power plants varies from cents 1.99/kWh (N3.16/kWh) to cents 5.65/kWh (N8.98/kWh).

5.1. Recommendations to improve performance of the selected power plants

To improve thermodynamic effectiveness of the selected gas turbine power plants, it is necessary to deal directly with inefficiencies related to exergy destruction and exergy loss in the systems. The primary contributors to exergy destruction are chemical reaction, heat transfer, mixing, and friction, including unrestrained expansions of gases and liquids. To deal with them effectively, the principal sources of inefficiency not only should be understood qualitatively, but also be determined quantitatively, at least approximately. Design changes to improve effectiveness must be done judiciously.

Based on the results of this research, the following possible technologies to improve performance of the selected gas turbine power plants are hereby recommended:

- The results of this study revealed that the combustion chamber has the largest irreversibility and cost of exergy destruction. This large exerge loss can be reduced in the selected power plants by addition of spray water and preheating of the reactants in the combustion chamber.

- Heat recovery from hot exhaust gases can be used to augment the performance of the gas turbine plant. Combined cycle is a common way to recover thermal energy from the exhaust gases; it is suitable for these plants as they operate as the base load plants (Oyedepo, Fagbenle, Adefila, & Adavbiele, 2014).

- Though gas turbine engines have the advantage of fast startup, they suffer from low power output and thermal efficiency at high ambient temperatures. GT power plants operating in Nigeria are simple GTs; there is a tremendous derating factor due to higher ambient temperatures. The average efficiency of GT plants in the Nigerian energy utility sector over the past two decades was in the range 27–30% (Abam, Ugot, & Igbong, 2012). Therefore, retrofitting GT power plants in Nigeria with advanced cycle would improve their performance significantly. Among many proven technologies are inlet air cooling, intercooling, regeneration, reheating, steam injection gas turbine etc. Air inlet cooling system (evaporative cooling, inlet fogging or inlet chilling method) is a useful option for increasing power output of the selected power plants. This helps to increase the density of the inlet air to the compressor.

- The compressor airfoils of older turbines tend to be rougher than a newer model simply because of longer exposure to the environment. In addition, the compressor of older models consumes a larger fraction of the power produced by the turbine section. Therefore, improving the performance of the compressor will have a proportionately greater impact on total engine performance. Application of Coatings to gas turbine compressor blades (the "cold end" of the machine) would improve the selected gas turbine engines performance. Compressor blade coatings provide smoother, more aerodynamic surfaces, which increase compressor efficiency. In addition, smoother surfaces tend to resist fouling because there are fewer "nooks and crannies" where dirt particles can attach. Coatings are designed to resist corrosion, which can be a significant source of performance degradation, particularly if a turbine is located near saltwater. As AES Barge gas turbine plant is located on lagoon, compressor coating technology would improve the plant performance significantly.

- Another option for improving the selected gas turbine plants performance is to apply ceramic coatings to internal components. Thermal barrier coatings (TBCs) are applied to hot section parts in advanced gas turbines. As some of the selected gas turbines are over 25 years in operation, TBCs can be applied to the hot sections of the selected gas turbines. The TBCs provide an insulating barrier between the hot combustion gases and the metal parts. TBCs will provide longer parts life at the same firing temperature, or will allow the user to increase firing temperature while maintaining the original design life of the hot section.

Nomenclature

Symbols

$\dot{E}$	exergy rate (kW)
$\dot{E}_L$	exergy loss rate
$\dot{E}_D$	exergy destruction rate
$ExIP$	exergetic improvement potential
h	specific enthalpy (kJ/kg)
I	irreversibility
ke	kinetic energy (kJ)
m	mass (kg)
$\dot{m}$	mass flow (kg/s)
p	pressure (bar)
P	power output (kW)
pe	potential energy (kJ)
$\dot{Q}$	heat (W)
r_p	pressure compression ratio
R	gas constant (kJ/mol K)
$\dot{S}$	entropy rate
T	temperature, either (K) or (°C)
T_{pz}	primary zone combustion temperature
$\dot{W}_c$	compressor work (W)
$\dot{W}_T$	turbine work (W)
$\dot{W}$	work (W)
y_D	exergy destruction rate ratio

Greek symbols

η_c	isentropic efficiency of compressor
η_T	isentropic efficiency of turbine
ηth	thermal efficiency
ε	exergetic efficiency
$\varnothing$	rational efficiency
δ	component efficiency defect
ψ	overall exergetic efficiency
ξ	exergetic performance coefficient

Subscripts

i	inlet
e	exit or outlet
p	pressure

a	air
pg	combustion product
f	fuel
T	turbine
cc	combustion chamber
th	thermal
sys	system
0	ambient
cv	control volume
D	destruction
gen	generation
ac	air compressor
gt	gas turbine
k	component

Superscripts

tot	total
PH	physical
KN	kinetic
PT	potential
CHE	chemical
T	thermal
P	mechanical

Abbreviation

LHV	lower heating value
TET	turbine exit temperature (K)
TDI	thermal discharge index
GTIT	gas turbine inlet temperature (K)

Acknowledgements
The authors appreciate the Management of AES power plant, Afam power plant and Ughelli power plant for providing the data used in this study.

Funding
Alam wishes to acknowledge supports given to him from the Research Grant Council of Shenzhen Government [grant number KQCX2014052114423867], [grant number JCYJ20130402100505796] and [grant number JCYJ20120613145300404].

Author details
S.O. Oyedepo[1]
E-mail: Sunday.oyedepo@covenantuniversity.edu.ng
R.O. Fagbenle[2]
E-mail: layifagbenle@yahoo.com
S.S. Adefila[3]
E-mail: samadefila@gmail.com
Md. Mahbub Alam[4]
E-mail: alamm28@yahoo.com
[1] Mechanical Engineering Department, Covenant University, Ota, Nigeria.
[2] Mechanical Engineering Department, Obafemi Awolowo University, Ile, Ife, Nigeria.
[3] Chemical Engineering Department, Covenant University, Ota, Nigeria.
[4] Institute for Turbulence-Noise-Vibration Interaction and Control, Shenzhen Graduate School, Harbin Institute of Technology, Shenzhen, China.

References
Abam, D. P. S., & Moses, N. N. (2011). Computer simulation of a gas turbine performance. *Global Journal of Research Engineering, 11*, 37–43.
Abam, F. I., Ugot, I. U., & Igbong, D. I. (2011). Thermodynamic assessment of grid-based gas turbine power plants in Nigeria. *Journal of Emerging Trends in Engineering and Applied Sciences, 2*, 1026–1033.
Abam, F. I., Ugot, I. U., & Igbong, D. I. (2012). Performance analysis and components irreversiblities of a (25 MW) gas turbine power plant modeled with a spray cooler.

American Journal of Engineering and Applied Sciences, 5, 35–41.

Ahmadi, P., Ameri, M., & Hamidi, A. (2009). Energy, exergy and exergoeconomic analysis of a steam power plant (a case study). *International Journal of Energy Research, 33,* 499–512.

Ahmadi, P., Barzegar, A. H., Ghaffarizadeh, A., & Saidi, M. H. (2010). Thermoeconomic-environmental multi-objective optimization of a gas turbine power plant with preheater using evolutionary algorithm. *International Journal of Energy Research, 35,* 389–403.

Ahmadi, P., & Dincer, I. (2011). Thermodynamic and exergoenvironmental analyses, and multi-objective optimization of a gas turbine power plant. *Applied Thermal Engineering, 31,* 14–15.

Ahmadi, P., Dincer, I., & Rosen, M. A. (2011). Exergy, exergoeconomic and environmental analyses and evolutionary algorithm based multi-objective optimization of combined cycle power plants. *Energy, 36,* 5886–5898. http://dx.doi.org/10.1016/j.energy.2011.08.034

Ahmadi, P., Rosen, M. A., & Dincer, I. (2011). Greenhouse gas emission and exergo-environmental analyses of a trigeneration energy system. *International Journal of Greenhouse Gas Control, 5,* 1540–1549. http://dx.doi.org/10.1016/j.ijggc.2011.08.011

Altayib, K. (2011). *Energy, exergy and exergoeconomic analyses of gas-turbine based systems* (MSc thesis). University of Ontario Institute of Technology, Oshawa.

Ameri, M., Ahmadi, P., & Hamidi, A. (2009). Energy, exergy and exergoeconomic analysis of a steam power plant: A case study. *International Journal of Energy Research, 33,* 499–512. http://dx.doi.org/10.1002/er.v33:5

Ameri, M., Ahmadi, P., & Khanmohammadi, S. (2008). Exergy analysis of a 420 MW combined cycle power plant. *International Journal of Energy Research, 32,* 175–183. http://dx.doi.org/10.1002/(ISSN)1099-114X

Aras, H., & Balli, O. (2008). Exergoeconomic analysis of a combined heat and power system with the micro gas turbine (MGTCHP). *Energy Exploration & Exploitation, 26,* 53–70.

Barzegar Avval, H., Ahmadi, P., Ghaffarizadeh, A., & Saidi, M. H. (2011). Thermo-economic-environmental multiobjective optimization of a gas turbine power plant with preheater using evolutionary algorithm. *International Journal of Energy Research, 35,* 389–403. http://dx.doi.org/10.1002/er.v35.5

Bejan, A., Tsatsaronis, G., & Moran, M. (1996). *Thermal design and optimization.* New York, NY: Wiley.

Can, O. F., Celik, N., & Dagtekin, I. (2009). Energetic–exergetic–economic analyses of a cogeneration thermic power plant in Turkey. *International Communications in Heat and Mass Transfer, 36,* 1044–1049. http://dx.doi.org/10.1016/j.icheatmasstransfer.2009.06.022

Demirel, Y. (2013). Thermodynamic analysis. *Arabian Journal for Science and Engineering, 38,* 221–249. http://dx.doi.org/10.1007/s13369-012-0450-8

Dincer, I., & Rosen, M. A. (2003). Thermoeconomic analysis of power plants: An application to a coal-fired electrical generating station. *Energy Conversion and Management, 44,* 2743–2761.

Fellah, G. M., Mgherbi, F. A., & Aboghres, S. M. (2010). Exergoeconomic analysis for unit Gt14 of South Tripoli gas turbine power plant. *Jordan Journal of Mechanical and Industrial Engineering, 4,* 507–516.

Gorji-Bandpy, M., & Ebrahimian, V. (2006). Exergoeconomic analysis of gas turbine power plant. *International Energy Journal, 7,* 37–41.

Gorji-Bandpy, M., & Goodarzian, H. (2011). Exergoeconomic optimization of gas turbine power plants operating

parameters using genetic algorithms: A case study. *Thermal Science, 15,* 43–54. http://dx.doi.org/10.2298/TSCI101108010G

Gorji-Bandpy, M., Goodarzian, H., & Biglari, M. (2010). The cost-effective analysis of a gas turbine power plant. *Energy Sources, Part B: Economics, Planning, and Policy, 5,* 348–358. http://dx.doi.org/10.1080/15567240903096894

Ibrahim bin, H., Masrul, H. M., Mohd Zamri, Y., & Mobd, H. B. (2001, August 6–8). Exergy costing for thermal plant optimization. In *Proceedings of the 3rd TNB Technical Conference.* Kajang-Puchong: UN/TEN.

Igbong, D. L., & Fakorede, D. O. (2014). Exergoeconomic analysis of a 100 MW unit GE Frame 9 gas turbine plant in Ughelli, Nigeria. *International Journal of Engineering and Technology, 4,* 463–468.

Kaviri, A. G., Mohd Jafar, M. N., Tholudin, M. L., & Barzegar Avval, H. (2011). Thermodynamic modeling and exergoeconomic optimization of a steam power plant using a genetic algorithm. *International Journal of Chemical and Environmental Engineering, 2,* 377–383.

Kim, S., Oh, S., Kwon, Y., & Kwak, H. (1998). Exergoeconomic analysis of thermal systems. *Energy, 23,* 393–406. http://dx.doi.org/10.1016/S0360-5442(97)00096-0

Kotas, T. J. (1995). *The exergy method of thermal plant analysis.* Malabar, FL: Krieger.

Kwon, Y., Kwak, H., & Oh, S. (2001). Exergoeconomic analysis of gas turbine cogeneration systems. *Exergy, An International Journal, 1,* 31–40. http://dx.doi.org/10.1016/S1164-0235(01)00007-3

Lazzaretto, A., & Tsatsaronis, G. (1999). On the calculation of efficiencies and costs in thermal systems. In S. M. Aceves, S. Garimella, & R. Peterson (Eds.), *Proceedings of the ASME Advanced Energy Systems Division, AES* (Vol. 39, pp. 421–430). New York, NY: ASME.

Marzouk, A., & Hanafib, A. (2013). Thermo-economic analysis of inlet air cooling in gas turbine plants. *Journal of Power Technologies, 93,* 90–99.

Moran, M. J. (1982). *Availability analysis: A guide to efficient energy use.* Englewood Cliffs, NJ: Prentice-Hall.

Moran, M. J., & Tsatsaronis, G. (2000). Engineering thermodynamic. In F. Kreith (Ed.), *The CRC handbook of thermal engineering.* Boca Raton, FL: CRC Press LLC.

Mousafarash, A., & Ahmadi, P. (2014). Exergy and exergo-economic based analysis of a gas turbine power generation system. In I. Dincer, A. Midilli, & H. Kucuk (Eds.), *Progress in sustainable energy technologies, Vol. II: Creating sustainable development* (pp. 1–12). New York, NY: Springer International.

Mousafarash, A., & Ameri, M. (2013). Exergy and exergo-economic based analysis of a gas turbine power generation system. *Journal of Power Technologies, 93,* 44–51.

Ofodu, J. C., & Abam, D. P. S. (2002). Exergy analysis of Afam thermal power plant. *NSE Technical Transactions, 37,* 14–28.

Oyedepo, S. O. (2014). *Thermodynamic performance analysis of selected gas turbine power plants in Nigeria* (PhD thesis). Ota: Covenant University.

Oyedepo, S. O., Fagbenle, R. O., Adefila, S. S., & Adavbiele, S. A. (2014). Performance evaluation and economic analysis of a gas turbine power plant in Nigeria. *Energy Conversion and Management, 79,* 431–440. http://dx.doi.org/10.1016/j.enconman.2013.12.034

Querol, E., Gonzalez-Regueral, B., & Perez-Benedito, J. L. (2013). *Practical approach to exergy and thermoeconomic analyses of industrial processes.* London: Springer. http://dx.doi.org/10.1007/978-1-4471-4622-3

Ray, T. T., Ganguly, R., & Gupta, A. (2007, July 16–20). Exergy analysis for performance optimization of a steam turbine cycle. In *IEEE PES Power African 2007 Conference and*

Exposition. Johannesburg. http://dx.doi.org/10.1109/PESAFR.2007.4498116

Rosen, M. A. (2009). Applications of exergy to enhance ecological and environmental understanding and stewardship. In *Proceedings of the 4th IASME / WSEAS International Conference on Energy & Environment (EE'09)* (pp. 146–152), Oshawa.

Sahoo, P. K. (2008). Exergoeconomic analysis and optimization of a cogeneration system using evolutionary programming. *Applied Thermal Engineering, 28*, 1580–1588. http://dx.doi.org/10.1016/j.applthermaleng.2007.10.011

Seyyedi, S. M., Ajam, H., & Farahat, S. (2010). A new iterative approach to the optimization of thermal energy systems: Application to the regenerative Brayton cycle. *Proceedings of the Institution of Mechanical Engineers, Part A: Journal of Power and Energy, 224*, 313–327.

Siahaya, Y. (2009). Thermoeconomic analysis of gas turbine power plant (GE MS 6001B PLTG – PLN – Sektor Tello Makassar). *Journal Penelitan Enjinring, 12*, 141–150.

Silvio, O. (2013). *Exergy: Production, cost and renewability*. London: Springer-Verlag.

Singh, O. K., & Kaushik, S. C. (2014). Exergoeconomic analysis of a Kalina cycle coupled coal-fired steam power plant. *International Journal of Exergy, 14*, 38–59. http://dx.doi.org/10.1504/IJEX.2014.059512

Tsatsaronis, G. (1993). Thermoeconomic analysis and optimization of energy systems. *Progress in Energy and Combustion Science, 19*, 227–257. http://dx.doi.org/10.1016/0360-1285(93)90016-8

Tsatsaronis, G., & Winhold, M. (1984). *Thermoeconomic analysis of power plants* (EPRI Final Report AP – 3651). Palo Alto, CA.

Valero, A., Lozano, M. A., Serra, L., Tsatsaronis, G., Pisa, J., Frangopoulos, C., & von Spakovsky, M. R. (1994). CGAM problem: Definition and conventional solution. *Energy, 19*, 279–286. http://dx.doi.org/10.1016/0360-5442(94)90112-0

11

A comparative study between chemical and enzymatic transesterification of high free fatty acid contained rubber seed oil for biodiesel production

Jilse Sebastian[1]*, Chandrasekharan Muraleedharan[1] and Arockiasamy Santhiagu[2]

*Corresponding author: Jilse Sebastian, Department of Mechanical Engineering, National Institute of Technology Calicut, Kozhikode 673601, Kerala, India

E-mail: jilsesebastian@gmail.com

Reviewing editor: An-Ping Zeng, Technische Universität Hamburg-Harburg, Germany

Abstract: The choice of a paramount method for biodiesel production has significance as the demand of alternative fuels like biodiesel is growing rapidly. In the present study, experimental results from chemical-catalysed as well as enzyme-catalysed methods were compared using common influencing parameters such as oil/alcohol molar ratio, catalyst concentration and reaction duration. Requirement of certain solvents to enhance the reaction rate was explained in the enzyme-catalysed transesterification reaction. Biodiesel conversion of more than 90% was attained for chemical-catalysed transesterification, whereas the conversion rate was 85% for enzyme-catalysed method. This gives the indication of further refinement in the enzyme-catalysed transesterification process. The influencing parameters and absolute results of the analysis give the impression of superiority of enzymatic transesterification method for biodiesel production from high free fatty acid-contained rubber seed oil.

ABOUT THE AUTHORS

Jilse Sebastian completed his master's degree in Internal Combustion Engineering from Anna University, Chennai. Currently, he is a doctoral research fellow at the Department of Mechanical Engineering, National Institute of Technology Calicut, Kerala, India, working in the field of biofuels.

Chandrasekharan Muraleedharan is a professor of Mechanical Engineering Department at National Institute of Technology Calicut, Kerala, India. He received his MSc (Engg.) and PhD from University of Calicut. He has more than 100 international research publications to his credit. His research interests include Fuels & Combustion, Environmental Pollution, Heat pipes and Energy Management.

Arockiasamy Santhiagu is working as an associate professor in School of Biotechnology at National Institute of Technology Calicut, Kerala, India. He received his M Tech and PhD from School of Biochemical Engineering Institute of Technology, Banaras Hindu University, Varanasi, India. He has more than 13 international research papers to his credit. His research interest includes Microbial production of enzymes, Bioremediation and Modelling and simulation of bioprocesses.

PUBLIC INTEREST STATEMENT

Development of a suitable method for biodiesel production is a needy matter in the present scenario. Transesterification of non-edible oils with a suitable catalyst is the most promising method for biodiesel production. The present study makes an experimental comparison between chemical transesterification and enzymatic transesterification. Enzymatic transesterification is a comparatively novel method for production of biodiesel from high free fatty acid-contained oils. The study describes the procedure of both the methods in detail and has made a comparison of experimental results.

Subjects: Automotive Technology & Engineering; Energy & Fuels; Mechanical Engineering; Power & Energy; Renewable Energy

Keywords: biodiesel; rubber seed oil; transesterification; enzyme

1. Introduction

Production of biodiesel to replace petroleum-refined diesel for the purpose of reduced pollution and sustainability is not a novel technique. The most established method for biodiesel production is chemically catalysed transesterification. But the cost of production and complexity in the processes lead to the need of alternate methods for biodiesel production. Enzyme-catalysed transesterification is a relatively new method for biodiesel production. In this method, enzymes which have the capability to catalyse transesterification of oils/fats (lipids) commonly known as lipases are used.

Chemical transesterification depends highly on the purity of the reactant oil especially free fatty acid (FFA) content. Canakci & Van Gerpen, 2001 reported from their experiments that in most of the cases, transesterification reaction stops at an acid value above 2 mgKOH/g while reacting on high FFA feed stocks. An acid esterification pre-treatment is mandatory for high FFA-contained vegetable oils. If the FFA content is too high, one-step acid esterification with sulphuric acid (H_2SO_4) may not be sufficient which may lead to two- or three-stage pre-treatment process. The pre-treated oil will then undergo alkaline transesterification for biodiesel production.

Enzymes are biological catalysts which accelerate the rate of a chemical reaction without undergoing a permanent change in the structure. Lipase is an enzyme that catalyses the breakdown or hydrolysis of fats (lipids) (Svendsen, 2000). Lipases fall in a subclass of esterases. Lipases (triacylglycerol acylhydrolases, EC 3.1.1.3) constitute various families of enzymes which are produced by animals, plants and micro-organisms. Since their bulk production is easier, commercialisation of microbial lipases and their involvement in enzymatic biodiesel production are more common than animal and plant ones. The microbes that have been suggested for biodiesel production include *Aspergillus niger, Candida cylindracea, Candida rugosa, Candida antarctica, Chromobacterium viscosum, Pseudomonas cepacia, Pseudomonas fluorescens, Rhizopus oryzae.* (Gog, Roman, Toşa, Paizs, & Irimie, 2012). Among these, *C. antarctica* displayed highest activity in methanolysis and ethanolysis reactions. *C. antarctica* in solvent-free environment resulted in above 90% conversion in majority of literature (Gog et al., 2012; Ho, Ngoc, Hyun, Mi, & Koo, 2007; Luque & Cervero, 2014; Modi, Reddy, Rao, & Prasad, 2007; Tupu, Jae, Marquis, Adesina, & Rogers, 2013).

Since lipases are insensitive to FFA, the oil does not require a pre-treatment and both the esterification and transesterification take place simultaneously to produce biodiesel directly. In addition to this, moisture content in the raw oil does not degrade the reaction, in fact will enhance the reactivity of the enzyme. However, the extended period of reaction and cost of enzymes restrict the wide acceptance of enzymatic transesterification for biodiesel production. Lipase immobilisation technology provides a number of important benefits including: (a) enzyme reuse, (b) easiness in separation of product from enzyme and (c) decrease in the inhibition rate. The process parameters are optimised in enzymatic transesterification process to make it a viable method for biodiesel production.

2. Materials and equipment

Crude rubber seed oil (RSO) was purchased from Pavalm & Co, Virudhunagar, Tamil Nadu, where the oil was extracted by expelling cured (de-shelled and dried) rubber seeds. Analytical grade methanol was used for transesterification. The bio-catalyst (enzyme) used was *C. antarctica lipase B* (Fermase CALB 10,000). Twenty-five grams of the lipase was supplied by Fermenta Biotech Ltd. Thane free of cost. PLU activity of the enzyme was specified as NLT (not less than) 10,000 U/g dry (One PLU enzyme unit corresponds to synthesis of 1 µmol of propyllaurate per minute per gram of immobilised enzyme at 60°C from lauric acid and 1-propanol). The enzyme was immobilised on polyacrylate beads and appeared as white- to creamish-white-coloured granular form.

Toluene, Isopropyl alcohol, n-Hexane, t-Butanol, Chloroform (all analytical grade) and deionised water were used as reagents for FFA estimation of oil and purification of biodiesel. A magnetic stirrer with hot plate was used as the mixing and heating equipment for chemical transesterification. The temperature was measured continuously with mercury column thermometer during the process.

In enzymatic transesterification, the reaction mixture was shaken in an incubated shaker in order to keep the stability of enzyme in the immobilisation medium. The incubator temperature was maintained as constant at 37°C throughout the experiments. The fatty acid compositions of RSO as well as the final biodiesel production yield were analysed by GC–MS technique in Care Keralam Ltd., Koratty, Kerala. The molecular formula of RSO was arrived as $C_{18}H_{32}O_{32}$ and molecular weight as 278. The properties of the oil are shown in Table 1.

The other equipment and instruments used include Centrifuge, Micro-Weighing balance, Micro pipette, Brookfield viscometer, Bomb calorimeter, Pensky Martin apparatus and Thermometers.

3. Chemical transesterification

In transesterification, reaction oil and an alcohol react together in which the glycerol backbone from the triglyceride molecule will be separated. The separated fatty acids will undergo reaction with alcohol to become a mixture of fatty acid alkyl esters which is termed as biodiesel. This reaction requires an elevated temperature of about 300°C with increased pressure (to avoid boiling of alcohol). Chemical catalysts will reduce the reaction temperature to a range of 50–60°C at normal atmospheric pressure condition. The commonly used chemical catalysts are potassium hydroxide (KOH) and sodium hydroxide (NaOH) .

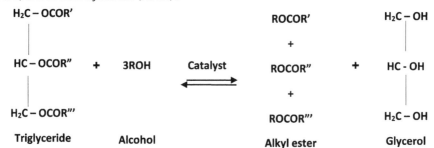

3.1. Influence of free fatty acids

FFAs are the chain-structured molecules which are separated from the glycerol backbone and usually have even number of carbon atoms. These FFAs do not take part in transesterification reaction and will make intermediate compound such as soap which has the capability to inhibit the biodiesel production reaction. The saponification characteristics of raw oil make it impractical to use in

Table 1. Properties of rubber seed oil	
Property	**Value**
Fatty acid composition (%)	
(1) Palmitic acid $C_{16:0}$	10.2
(2) Stearic acid $C_{18:0}$	8.7
(3) Oleic acid $C_{18:1}$	24.6
(4) Linoleic acid $C_{18:2}$	39.6
(5) Linolenic acid $C_{18:3}$	16.3
Specific gravity	0.91
Viscosity (mm²/s) at 30°C	59.7
Flash point (°C)	210
Acid value (mgKOH/g)	82

biodiesel production. The researchers through their vast experience claim that FFA content less than 2% is indeed to go for the transesterification reaction with alkali catalysts (Bharathiraja et al., 2014; Ramadhas, Jayaraj, & Muraleedharan, 2005). Therefore, a pre-treatment process (acid esterification) should be done on the raw oil before proceeding to alkali transesterification for biodiesel production.

3.2. Acid esterification

Standard AOCS titration method (Lubrizol, 2013) was adopted to determine acid value of the oil (percentage of FFA is half the acid value). In acid esterification, the FFA will react to alcohol (methanol) in the presence of sulphuric acid (H_2SO_4) so that it will be converted into alkyl ester and water.

$$FFA + Methanol \xrightarrow{Acid\ catalyst} Methyl\ ester + Water$$

FFA content of RSO was very high (42%) so that one-stage esterification was insufficient to reduce it to the desired level. Canakci and Van Gerpen (2001) reported from their experiments that in most of the cases, acid-catalysed reaction stops at an acid value above 2 mgKOH/g while reacting on high FFA feed stocks. Two-step acid esterification technique was used in the present study to reduce the acid value to the desirable range.

Different sets of experiments were conducted with varying oil to alcohol molar ratio and acid catalyst concentration. In each run, 200 ml of RSO was filled in the flask of 500-ml capacity and pre-heated to 50°C. Pre-determined amount of methanol was then added to the oil and stirring was started. After a few minutes, acid catalyst was added to the mixture and the stirring was continued. In the first step, reaction duration was 1 h, keeping the temperature at 45 ± 5°C throughout and the acid catalyst concentration at 1% (v/v). After the reaction, the mixture was kept in a separating funnel for 24 h. Water and methanol mixture along with sulphuric acid content moved to top. There was a layer of residue settled at the bottom containing colouring pigments, gum forming agents, wax, etc.

The second stage of acid esterification was carried out with different molar ratios keeping other parameters same as in the first stage. Three-stage reactions were also carried out with lower methanol ratio, and it was observed that the increase in methanol ratio caused a significant reduction in acid value. The residue and methanol water mixture were separated after the reaction. Five sets of selected experimental results are presented in Table 1. Use of an overall molar ratio of 80% v/v methanol and 2% v/v concentration of H_2SO_4 in two stages is found to give optimum results. The result obtained from pre-treatment process shows relatively less oil–methanol ratio in comparison with the work done by Satyanarayana and Muraleedharan (2010). Experimental trials were also carried out with varying acid catalyst concentration with optimum alcohol to oil molar ratio.

3.3. Effect of acid catalyst concentration

Ahmad, Yusup, Bokhari, and Kamil (2014) conducted experiments with 10 wt% H_2SO_4 in which 45% FFA-contained RSO was reduced to 0.82% in single-step acid treatment. Experiments were conducted with varying acid catalyst amount from 1% to 10% with favourable methanol ratio in the first-stage reaction. Even if the FFA content reduced continuously with increase in H_2SO_4, it did not reach to the desired value (2%) even at 10% v/v (refer Figure 1). Saponification occurred up to a FFA content of 3% on water washing. Therefore, multistage acid esterification was suggested as preferred. The catalyst concentration of 1% w/w was recommended for the first-stage reaction after considering the molar ratio and number of reaction stages (see Table 2).

3.4. Effect of molar ratio

The amount of alcohol used in both acid esterification and alkaline transesterification plays a significant role in reaction rate and overall production cost of biodiesel. Experiments were carried out with different methanol to oil molar ratios of 0.20, 0.40, 0.50, 0.60 and 0.65. A significant reduction in acid value was observed by increasing the percentage of oil–alcohol ratio up to 0.50. The trend of FFA

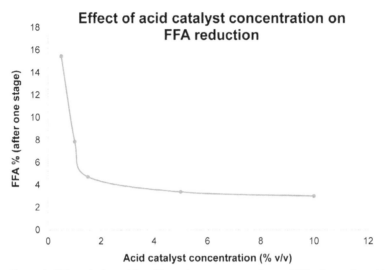

Figure 1. FFA variation with acid catalyst concentration at 50% v/v methanol in oil (after first-step pre-treatment).

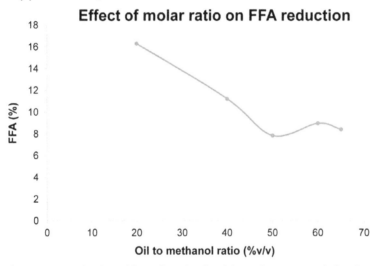

Figure 2. FFA reduction with methanol ratio with catalyst 1% w/w (after first-step pre-treatment).

Table 2. Absolute values of FFA content in the oil (temperature 45 ± 5°C, 1-h reaction, 1% v/v H_2SO_4 in each step)				
Set No.	Reaction stages	Molar ratio (% v/v)	Titration end point value (ml)	FFA (%)
1	1st stage	20	29	16.29
	2nd stage	20	15	8.41
	3rd stage	20	5	2.8
2	1st stage	40	20	11.22
	2nd stage	30	9	5.04
	3rd stage	20	3	1.68
3	1st stage	50	14	7.85
	2nd stage	30	3.6	2.02
4	1st stage	60	16	8.97
	2nd stage	20	4	2.44
5	1st stage	65	15	8.41
	2nd stage	20	4.5	2.52

reduction with molar ratio in first stage of pre-treatment is shown in Figure 2. It was noted that after the limit of 0.50, use of excess alcohol did not give any significant reduction of FFA content.

The second step of acid esterification was done for further reduction of FFA in the oil. Catalyst concentration was kept as 1% v/v for each set of second-step esterification. The molar ratios of each step with resultant FFA content are listed in Table 2.

4. Alkaline transesterification

Alkaline transesterification was carried out after the completion of two-stage pre-treatment process. The base catalyst used was KOH (85% assay) and molar ratio of oil to methanol was kept as 1:9 in each experiment. Properly mixed solution of KOH and methanol of required quantity was added to 150 ml of treated and preheated oil (50°C) and was stirred well. The temperature was maintained at 55 ± 3°C throughout the experiment for 40-min reaction time. The previous studies and literatures showed that there was no significant increase in biodiesel yield with increase in reaction time above 40 min (Satyanarayana & Muraleedharan, 2010). After the completion of the reaction, products were allowed to settle for 24 h. Glycerine was removed when it was settled at the bottom. The top layer (biodiesel) was purified by washing 3–5 times with demineralised water at room temperature.

4.1. Effect of base catalyst

Base catalyst concentration was varied from 0.5 wt% to 2 wt% and the result revealed that maximum yield of 90% was obtained with a base catalyst concentration of 1.5 wt% (refer Figure 3). The conversion efficiency did not improve after the limit of 1.5 wt% KOH addition. Emulsion was observed with lower catalyst concentrations (0.5–1 wt%) due to insufficient catalyst amount to completely neutralise the entire FFA resulting in saponification reaction. The soap content in the biodiesel made emulsion during water washing.

An excellent medium to break the emulsion was sodium chloride (NaCl) solution. NaCl solution of 50% by volume was added to the emulsion and was heated to 60°C and stirred. It has been observed that the biodiesel separated from the emulsion solution is coming out at the top. The entire chemical transesterification process including the two-step acid treatment and alkaline transesterification is summarised as a flow diagram shown in Figure 4.

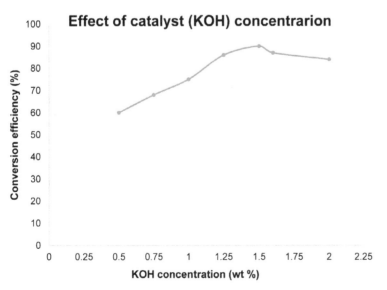

Figure 3. Effect of catalyst concentration on conversion efficiency at molar ratio 9:1 (methanol–oil).

Flow Chart

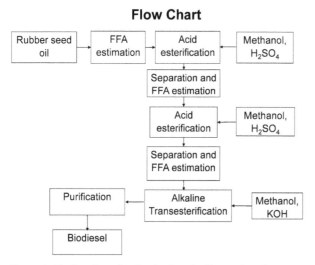

Figure 4. Biodiesel production in chemically catalysed transesterification (flow chart).

5. Enzymatic transesterification

The use of enzymes as catalysts for transesterification of high FFA-contained oils is a promising technique as the enzymes are insensitive to FFA. Pre-treatment process was eliminated in this method. Different sets of experiments were carried out to study the yield of biodiesel with different molar ratio, enzyme concentration, type of solvent and duration of reaction. RSO of 10 ml was used in each experimental run. Reactions were carried out in a conical flask of 100-ml working volume. The enzyme was weighed according to the wt% of oil in each set and added to the RSO. Methanol was measured by micropipette and added to the oil stepwise (three steps) at equal time interval. The reaction mixture with methanol, oil and enzyme was shaken in an incubated shaker at 170–200 rpm keeping the temperature at 37°C. After 24 h, reaction enzyme was separated out from the reaction mixture by filtration. In order to reduce time for experimental process, the mixture of biodiesel and glycerol produced after the reaction was centrifuged at 5,000 rpm for 15 min. Glycerol was settled at the bottom of the centrifuge tube and was removed. The separated mixture was heated above 100°C to remove unreacted methanol and water droplets. The enzyme separated was washed with distilled water and t-butanol, dried and stored under the required temperature (below 8°C) for further use.

5.1. Effect of methanol ratio

Influence of methanol/oil molar ratio was measured by varying it between 2 and 6. Even though the theoretical molar ratio required is 3, most of the researchers suggested a higher molar ratio in order to avoid the insufficiency of alcohol for the reaction (Gog et al., 2012; Yan et al., 2014). Sufficient alcohol may not be available for the reaction due to the traces of water and other liquid impurities in the oil which will be dissolved in it (Ramadhas et al., 2005). For each set of experiments, the enzyme concentration was maintained as 10% w/w of oil and no solvent was added to enhance the reaction rate. The trend of molar ratio against biodiesel conversion efficiency is plotted in Figure 5. Even if the conversion efficiency of molar ratio 6 was slightly higher than that of molar ratio 4, upon optimising considerations, molar ratio 4 was taken as recommended value.

5.2. Effect of enzyme concentration

The variation of product yield and kinematic viscosity was analysed with different enzyme concentrations. Enzyme weights between 5 and 12 were used in different sets by keeping the methanol/oil molar ratio at optimum value (molar ratio 4). According to Sanchez and Vasudevan (2006) and Chattopadhyay, Karemore, Das, Deysarkar, and Sen (2011), increase in reaction rate is directly proportional to enzyme concentration. In the present experiments, the biodiesel conversion ability was continuously increased but reduced beyond 10 wt%. The resulted viscosity value of the products is plotted in Figure 6. The excess solid enzyme available in the reactive mixture had the ability to limit

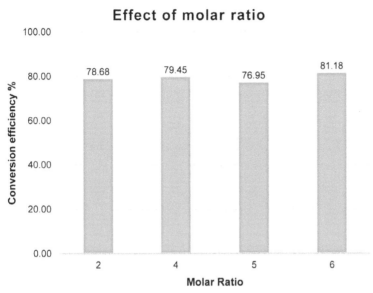

Figure 5. Effect of molar ratio on biodiesel conversion efficiency with 10% w/w enzyme concentration.

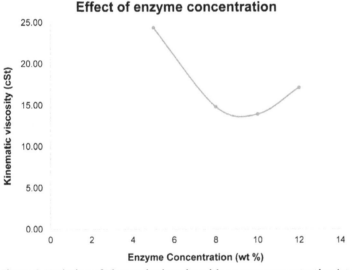

Figure 6. Variation of Kinematic viscosity with enzyme concentration (at molar ratio 4).

glycerol separation due to the non-interaction between methanol and oil. Similar result was reported by Shah and Gupta (2007) such that the maximum conversion was obtained at 10% w/w of oil.

5.3. Effect of solvent addition

Lower chain alcohols have the tendency to inhibit the reactivity of the lipases. The addition of solvents to the reaction mixture will enhance the enzyme activity. The solvents prevent the inhibition of enzyme activity by dissolving the alcohol and give the flexibility to the enzyme. In addition to this, glycerol produced during the early stages of transesterification has adverse effect on biodiesel production by accumulating on the immobilisation support (Moreno-Pirajan & Giraldo, 2011). Water and t-butanol were the commonly used solvents for hydrolysis reaction of oils and fats (Luque, Cervero, & Alvarez, 2014). According to Sun, Yu, Curran, and Liu (2012), lipases act efficiently at oil–water interface. Also t-butanol has greater ability to dissolve glycerol. Addition of small amount of water (not more than 5% w/w of oil) is preferable in lipase-catalysed transesterification (Gog et al., 2012; Yan et al., 2014). Different sets of experiments were done with and without solvent additions keeping other parameters optimum. The biodiesel conversion rate and kinematic viscosity of the products of

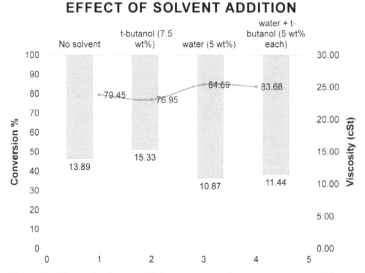

Figure 7. Effect of solvent addition on conversion percentage and kinematic viscosity (enzyme concentration 10 wt%, molar ratio 4).

reaction are expressed in the following graph (Figure 7). The absolute values of conversion percentage were mentioned in the line diagram. The downward-directed bar diagrams represent the reduction in viscosity value from the viscosity of raw oil and the absolute kinematic viscosity values were represented in cSt.

A maximum conversion percentage of 79.45 resulted in the experiments with no solvent addition. The addition of 7.5% w/w t-butanol has a reduced conversion rate, whereas the addition of 5% w/w of water significantly increased the conversion rate and reached to a maximum of 84.69%. Both the solvents were added in three steps at the time of methanol addition. A separate experiment was conducted by the addition of both the solvents each with 5% w/w of oil. The second solvent (t-butanol) was added after a time gap of 8 h from the beginning of reaction. The results showed that there was no improvement in conversion efficiency with addition of t-butanol. It can be inferred that the effect of enzyme activity inhibition by the presence of glycerol was insignificant in low-quantity enzymatic RSO biodiesel production. Therefore, the best solvent medium for enzyme transesterification for RSO was identified as water with methanol.

5.4. Effect of reaction duration

Since the production duration of biodiesel in enzyme-catalysed method is longer than chemical method, determination of optimum reaction duration is highly important in enzymatic transesterification. Experiments were performed with different durations of reaction. Reaction time was extended up to 43 h keeping other parameters optimum. It has been recorded that the maximum conversion is obtained for a duration of 30 h (Figure 8) after which the overall conversion percentage was less. This may be due to the reverse reaction in the transesterification process.

6. Statistical analysis of influencing factors for enzymatic transesterification

Since parameter optimisation was done based on literature reviews, statistical testing was avoided in chemical transesterification reaction method. Being a novel method of biodiesel production, one statistical model was formulated to study the influence of parameters on biodiesel production yield in enzymatic transesterification. The 2^3 full factorial model was adopted for the study. Only the concentration parameters were considered for analysis. The factors taken for analysis were enzyme concentration (E), molar ratio (M) and solvent addition (S). The software tool used was Design-Expert® Version 10. Both enzyme concentration and molar ratio were considered as numerical variables having limits 5, 10 wt% and 4, 5 v/v%, respectively. Solvent addition term was entered as categorical variable with "Yes" (5 wt%) and "No" (0 wt%). Table 3 shows the details of ANOVA.

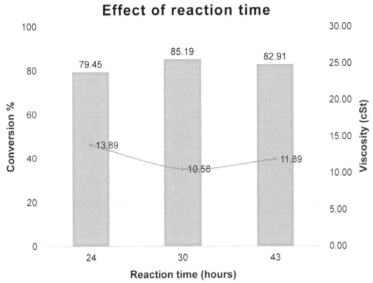

Figure 8. Effect of reaction duration on biodiesel conversion and kinematic viscosity (10 wt% enzyme, molar ratio 4 and 5% water).

Table 3. ANOVA table					
Source	**Sum of squares**	**Degree of freedom (df)**	**Mean square**	**F value**	**p-value prob > F**
Model	2.986E + 008	5	5.972E + 007	199.56	0.0050*
A-E	2.542E + 008	1	2.542E + 008	849.25	0.0012*
B-M	3.404E + 006	1	3.404E + 006	11.37	0.0778
C-S	3.636E + 007	1	3.636E + 007	121.49	0.0081*
AB	4.016E + 006	1	4.016E + 006	13.42	0.0671
BC	6.772E + 005	1	6.772E + 005	2.26	0.2714
Residual	5.985E + 005	2	2.993E + 005		

*p-values are significant.

From the table, p-value is less than 0.0500 which indicates that model terms are significant. In this case, E and S are significant model terms and M is insignificant. The table also infers that interactive effect of variables E-M and E-S is insignificant and E-M-S is too small enough to remove from normal plot. There is only 0.50% chance that an F-value could occur due to noise. The regression equation in terms of actual values at solvent added condition (S = Yes):

$$(Y)^{2.32} = -8679.24 + 4805.13 * E + 3528.34 * M - 566.8 * E * M$$

7. Comparison between chemical transesterification and enzymatic transesterification

Based on the experimental results, the important parameters on biodiesel production process through transesterification can be summarised (refer Table 2). It is important to recommend the suitable method for the biodiesel production from high FFA-contained RSO. Table 4 gives a comparison between chemical and enzymatic transesterification processes with respect to important aspects.

Other than reaction duration and catalyst concentration, all other parameters give superiority to enzymatic reaction. Use of immobilised enzyme catalyst can compensate higher catalyst consumption problem since the enzyme can be reused for several batch operations. Even though the chemical reaction duration is quite less in comparison with enzymatic reaction, the overall biodiesel

Sl. No.	Property	Chemical transesterification	Enzymatic transesterification
		Table 4. Comparison of important parameters in the transesterification reactions	
1	Catalyst	KOH	*Candida antarctica lipase B*
2	Oil/methanol molar ratio	1:9	1:4
3	Catalyst concentration	1% w/w	10% w/w
4	Sensitivity to FFA	Highly sensitive, direct reaction leads to saponification	Insensitive to FFA content in the oil
5	Necessity of pre-treatment	Two-stage acid esterification (pre-treatment) is needed to meet the required oil quality	No need of pre-treatment
6	Reaction period	2–3 h (including pre-treatment)	Minimum 24 h
7	Sensitivity to water content in oil	Highly sensitive, need to remove entire water particles before the start of reaction	Presence of water (up to 5% w/w) can improve the rate of reaction
8	Solvent addition	No solvent needed	Addition of solvents improves the reaction rate significantly
9	Biodiesel production yield	More than 90%	About 85%
10	Purity of by-product glycerol	Impure and needs further purification steps	Glycerol separated is pure

production period was comparable. This is due to the repeated water washing and separation steps needed due to high FFA content of oil in chemical transesterification method.

8. Conclusion

Enzymatic transesterification being the newly identified technique for biodiesel production needs further refinements to meet the replacement of the established method of chemical transesterification. On comparison of the two methods, it can be concluded that enzyme-catalysed reaction is superior to chemical method. Enzyme-catalysed transesterification has given more pure biodiesel as well as glycerol than that from chemical method. Still, the conversion percentage of enzyme-catalysed transesterification is less. The increased reaction duration of enzymatic transesterification restricts the production process to batch production method. High-level mixing methods such as ultrasonic vibration method can be adopted to increase the rate of reaction in enzymatic transesterification. However, upon absolute measurement of total process duration including pre-treatment stages and gravity separation, chemical transesterification production of high FFA-contained RSO is lengthier. The traces of highly reactive acid and base contents in the reaction products need several levels of water washing and neutralisation in chemical method, thereby increasing the number of downstream processes. The present experiments also throw light on identification of the potential of underutilised RSO feedstock for biodiesel production.

Funding
The authors received no direct funding for this research.

Author details
Jilse Sebastian[1]
E-mail: jilsesebastian@gmail.com
ORCID ID: http://orcid.org/0000-0002-2611-9654
Chandrasekharan Muraleedharan[1]
E-mail: murali@nitc.ac.in
Arockiasamy Santhiagu[2]
E-mail: asanthiagu@nitc.ac.in

[1] Department of Mechanical Engineering, National Institute of Technology Calicut, Kozhikode 673601, Kerala, India.
[2] School of Biotechnology, National Institute of Technology Calicut, Kozhikode 673601, Kerala, India.

References

Ahmad, J., Yusup, S., Bokhari, A., & Kamil, R. N. M. (2014). Study of fuel properties of rubber seed oil based biodiesel. *Energy Conversion and Management, 78*, 266–275. doi:10.1016/j.enconman.2013.10.056

Bharathiraja, B., Chakravarthy, M., Kumar, R. R., Yuvaraj, D., Jayamuthunagai, J., Kumar, R. P., & Palani, S. (2014). Biodiesel production using chemical and biological methods–A review of process, catalyst, acyl acceptor, source and process variables. *Renewable and Sustainable Energy Reviews, 38*, 368–382. doi:10.1016/j.rser.2014.05.084

Canakci, M., & Van Gerpen, J. (2001). Biodiesel production from oils and fats with high free fatty acids. *Transactions-American Society of Agricultural Engineers, 44*, 1429–1436.

Chattopadhyay, S., Karemore, A., Das, S., Deysarkar, A., & Sen, R. (2011). Biocatalytic production of biodiesel from cottonseed oil: Standardization of process parameters and comparison of fuel characteristics. *Applied Energy, 88*, 1251–1256. doi:10.1016/j.apenergy.2010.10.007

Gog, A., Roman, M., Toşa, M., Paizs, C., & Irimie, F. D. (2012). Biodiesel production using enzymatic transesterification–Current state and perspectives. *Renewable Energy, 39*, 10–16. doi:10.1016/j.renene.2011.08.007

Ho, S., Ngoc, M., Hyun, S., Mi, S., & Koo, Y. (2007). Lipase-catalyzed biodiesel production from soybean oil in ionic liquids. *Enzyme and Microbial Technology, 41*, 480–483. doi:10.1016/j.enzmictec.2007.03.017

Lubrizol. (2013). Determination of acid value and free fatty acid. *Procedure, Lubrizol test*. Retrieved from 2013, www.lubrizol.com/personalcare

Luque, S., Cervero, J. M., & Alvarez, J. R. (2014). Novozym 435-catalyzed synthesis of fatty acid ethyl esters from soybean oil for biodiesel production. *Biomass and Bioenergy, 61*, 131–137. doi:10.1016/j.biombioe.2013.12.005

Modi, M. K., Reddy, J. R. C., Rao, B. V. S. K., & Prasad, R. B. N. (2007). Lipase-mediated conversion of vegetable oils into biodiesel using ethyl acetate as acyl acceptor. *Bioresource Technology, 98*, 1260–1264. doi:10.1016/j.biortech.2006.05.006

Moreno-Pirajan, J. C., Giraldo, L. (2011). Study of immobilized *candida rugosa* lipase for biodiesel fuel production from palm oil by flow microcalorimetry. *Arabian Journal of Chemistry, 4*, 55–62. doi:10.1016/j.arabjc.2010.06.019

Ramadhas, A. S., Jayaraj, S., & Muraleedharan, C. (2005). Biodiesel production from high FFA rubber seed oil. *Fuel, 84*, 335–340. doi:10.1016/j.fuel.2004.09.016

Sanchez, F., & Vasudevan, T. P. (2006). Enzyme catalyzed production of biodiesel from olive oil. *Applied Biochemistry and Biotechnology, 135*, 1–14.

Satyanarayana, M., & Muraleedharan, C. (2010). Methyl ester production from rubber seed oil using two-step pretreatment process. *International Journal of Green Energy, 7*, 84–90. doi:10.1080/15435070903501290

Shah, S., & Gupta, M. N. (2007). Lipase catalyzed preparation of biodiesel from Jatropha oil in a solvent free system. *Process Biochemistry, 42*, 409–414. doi:10.1016/j.procbio.2006.09.024

Sun, J., Yu, B., Curran, P., & Liu, S. (2012). Lipase-catalysed transesterification of coconut oil with fusel alcohols in a solvent-free system. *Food Chemistry, 134*, 89–94. doi:10.1016/j.foodchem.2012.02.070

Svendsen, A. (2000). Lipase protein engineering. *Biochimica et Biophysica Acta, 1543*, 223–238. doi:10.1016/S0167-4838(00)00239-9

Tupu, S. C., Jae, Y., Marquis, C., Adesina, A. A., & Rogers, P. L. (2013). Enzymatic conversion of coconut oil for biodiesel production. *Fuel Processing Technology, 106*, 721–726. doi:10.1016/j.fuproc.2012.10.007

Yan, Y., Li, X., Wang, G., Gui, X., Li, G., Su, F., & Liu, T. (2014). Biotechnological preparation of biodiesel and its high-valued derivatives: A review. *Applied Energy, 113*, 1614–1631. doi:10.1016/j.apenergy.2013.09.029

12

Varied effects of shear correction on thermal vibration of functionally graded material shells

C.C. Hong[1*]

*Corresponding author: C.C. Hong, Department of Mechanical Engineering, Hsiuping University of Science and Technology, Taichung 412-80, Taiwan, ROC
E-mail: cchong@mail.hust.edu.tw
Reviewing editor: Duc Pham, University of Birmingham, UK

Abstract: The effects of varied shear correction coefficient on the first-order transverse shear deformation result of functionally graded material (FGM) thick circular cylindrical shells under thermal vibration are investigated and computed by using the generalized differential quadrature method. The computed and varied values of shear correction coefficient are usually functions of FGM power law index and environment temperature. In the thermoelastic stress–strain relations, the simpler form stiffness of FGM shells under linear temperature rise is considered. The equation of shear correction coefficient is derived and obtained by using the total strain energy principle. Two parametric effects: environment temperature and FGM power law index on the thermal stress and center deflection of FGM thick circular cylindrical shells are obtained and investigated.

Keywords: varied shear correction coefficient, FGM, shell, thermal vibration, GDQ

1. Introduction

There are some vibration researches of the functionally graded material (FGM) shells. Ebrahimi and Najafizadeh (2014) used the generalized differential quadrature (GDQ) and generalized integral quadrature methods to calculate the free vibration results for a two-dimensional FGM circular cylindrical shell with the Love's first approximation classical shell theory. It is often convenient to use GDQ method to study the vibration of FGM shells. Du, Li, and Jin (2014) used the Lagrangian theory and multiple scale method to investigate the forced vibration of infinitely long FGM cylindrical shells. Strozzi and Pellicano (2013) used the Sanders–Koiter theory to study the numerical non-linear vibrations of FGM circular cylindrical shells. The non-linear dynamic character of the shell geometry and material properties of FGM are studied in more detailed. Some shear deformation theories usually

ABOUT THE AUTHOR

My key research activities are applications of magnetostrictive actuator and permanent magnet actuator, generalized differential quadrature (GDQ) computations of magnetostrictive material (Terfenol-D), functionally graded material (FGM) plates, and shells under thermal vibration with/without shear deformation. In this paper, the effects of varied shear correction coefficient on the first-order transverse shear deformation result of functionally graded material (FGM) thick circular cylindrical shells under thermal vibration are investigated and computed by using the GDQ method.

PUBLIC INTEREST STATEMENT

The GDQ method is a computational method used to calculate the thermal stress and deflection of FGM shells under the environment temperature. The FGM is a functionally graded material usually has the properties are functions of environment temperature. For the thick shells, it is necessary to consider the first-order shear deformation theory (FSDT) for the displacements with middle-surface shear rotations. Usually, the value of shear correction coefficient used in the integration of shear stiffness is not constant in the FGM shells. The varied, modified shear correction factor for shear correction coefficient usually used and can be derived based on the energy equivalence principle.

applied in the calculations of shear resultants for FGM shells, there are the first-order shear deformation theory (FSDT) and the higher order shear deformation theory (HSDT). Sheng and Wang (2013a) presented the non-linear vibration control of thin piezoelectric layers embedded on inner and outer surfaces of FGM cylindrical shells by using the Hamilton's principle and Von Karman non-linear theory. Sheng and Wang (2013b) used the Hamilton's principle and Von Karman non-linear theory to investigate the vibrations of FGM cylindrical shells by considering the FSDT, thermal, and axial loads. Neves et al. (2013) applied the virtual work principle to study the free vibration of FGM shells with the HSDT. Malekzadeh and Heydarpour (2012) used the Hamilton's principle to obtain the differential quadrature method (DQM) vibration results for the rotating FGM cylindrical shells with FSDT under thermal environment. Malekzadeh, Fiouz, and Sobhrouyan (2012) used the DQM to calculate the free vibration solutions for the FGM truncated conical shells in thermal environment. Li, Fu, and Batra (2010) used Flugge's shell theory to present the computed results of free vibrations for the composited FGM circular cylindrical shell. Cinefra, Belouettar, Soave, and Carrera (2010) applied the variable kinematic shell model to present the free-vibration closed-form solutions of FGM shells. Usually, the constant correction factor value 5/6 was used in the theoretical shear stress calculation for the FGM shells did not consider with the material forms of symmetry and the shell thickness. Chróścielewski, Pietraszkiewicz, and Witkowski (2010) presented some influence examples of the different correction factor values on the highly non-linear shell structures. For the non-symmetrical and thick thickness of FGM shells, the implantation of calculated and varied shear correction is needed to be considered.

The author has some GDQ computational experiences in the thermal vibration and transient response of composited shells. Hong (2013a) investigated the rapid heating induced vibration of Terfenol-D FGM circular cylindrical shells without considering the shear deformation effects. Hong (2010) studied the computational approach of piezoelectric shells by considering the shear deformation effects. The actual and practical use of such FGM shells theoretical studies might be applied in the field of structure where the stress and deformation becomes sensitive when they undergo thermal vibration. It is interesting in this FSDT with the varied effects of shear correction coefficient of FGM thick circular cylindrical shells with four edges in simply supported boundary conditions, thermal stresses, and center deflection of GDQ computation are obtained. Two parametric effects of environment temperature and FGM power law index on the thermal stress and center deflection of FGM thick circular cylindrical shells are also investigated.

2. Formulation

Two-layered FGM circular cylindrical shell is shown in Figure 1 with FGM layer 1 thickness h_1 and FGM layer 2 thickness h_2, respectively. The material properties of power-law function of FGM circular

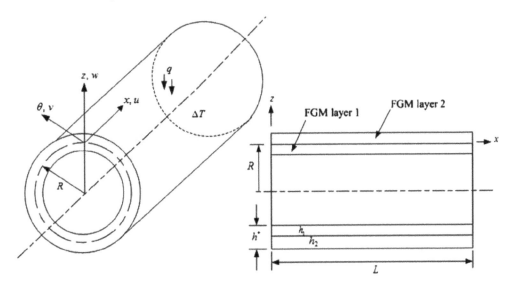

Figure 1. Two-layered FGM circular cylindrical shell.

cylindrical shells are considered with Young's modulus E_{fgm} of FGM in standard variation form of power law index R_n, the others are assumed in the simple average form (Chi & Chung, 2006; Hong, 2013b). The properties P_i of individual constituent material of FGMs are functions of environment temperature T in the following form (Hong, 2012).

$$P_i = P_0(P_{-1}T^{-1} + 1 + P_1T + P_2T^2 + P_3T^3)$$

(1)

where P_0, P_{-1}, P_1, P_2, and P_3 are the temperature coefficients.

The time dependent of displacements u, v, and w of thick circular cylindrical shells are assumed in the following linear FSDT equations (Qatu, Sullivan, & Wang, 2010).

$$u = u_0(x, \theta, t) + z\varphi_x(x, \theta, t)$$

(2a)

$$v = v_0(x, \theta, t) + z\varphi_\theta(x, \theta, t)$$

(2b)

$$w = w(x, \theta, t)$$

(2c)

where u_0 and v_0 are tangential displacements, w is transverse displacement of the middle-surface of the shells, φ_x and φ_θ are middle-surface shear rotations, x and θ are in-surface coordinates of the shell, z is out of surface coordinates of the shell, and t is time.

For the plane stresses in the thick FGM circular cylindrical shell, the in-plane stresses constitute the membrane stresses, bending stresses, and thermal stresses under temperature difference ΔT for the kth layer are in the following equations (Lee, Reddy, & Rostam-Abadi, 2006; Whitney, 1987).

$$\left\{ \begin{array}{c} \sigma_x \\ \sigma_\theta \\ \sigma_{x\theta} \end{array} \right\}_{(k)} = \left[\begin{array}{ccc} \bar{Q}_{11} & \bar{Q}_{12} & \bar{Q}_{16} \\ \bar{Q}_{12} & \bar{Q}_{22} & \bar{Q}_{26} \\ \bar{Q}_{16} & \bar{Q}_{26} & \bar{Q}_{66} \end{array} \right]_{(k)} \left\{ \begin{array}{c} \varepsilon_x - \alpha_x \Delta T \\ \varepsilon_\theta - \alpha_\theta \Delta T \\ \varepsilon_{x\theta} - \alpha_{x\theta} \Delta T \end{array} \right\}_{(k)}$$

(3a)

and the shear stresses are given as follows.

$$\left\{ \begin{array}{c} \sigma_{\theta z} \\ \sigma_{xz} \end{array} \right\}_{(k)} = \left[\begin{array}{cc} \bar{Q}_{44} & \bar{Q}_{45} \\ \bar{Q}_{45} & \bar{Q}_{55} \end{array} \right]_{(k)} \left\{ \begin{array}{c} \varepsilon_{\theta z} \\ \varepsilon_{xz} \end{array} \right\}_{(k)}$$

(3b)

where α_x and α_θ are the coefficients of thermal expansion, $\alpha_{x\theta}$ is the coefficient of thermal shear, $\bar{Q}_{ij}$ is the stiffness of FGM shell. ε_x, ε_θ, and $\varepsilon_{x\theta}$ are in-plane strains, not negligible $\varepsilon_{\theta z}$ and ε_{xz} are shear strains, the curvatures of k_x, k_θ, and $k_{x\theta}$, respectively, in terms of displacement components and shear rotation are given as follows.

$$\varepsilon_x = \frac{\partial u_0}{\partial x} + z\frac{\partial \varphi_x}{\partial x}$$

(4a)

$$\varepsilon_\theta = \frac{1}{R}\left(\frac{\partial v_0}{\partial \theta} + z\frac{\partial \varphi_\theta}{\partial \theta} + w \right)$$

(4b)

$$\varepsilon_{x\theta} = \frac{\partial v_0}{\partial x} + z\frac{\partial \varphi_\theta}{\partial x} + \frac{1}{R}\left(\frac{\partial u_0}{\partial \theta} + z\frac{\partial \varphi_x}{\partial \theta} \right)$$

(4c)

$$\varepsilon_{\theta z} = \varphi_\theta + \frac{1}{R}\frac{\partial w}{\partial \theta} - \frac{v_0}{R}$$

(4d)

$$\varepsilon_{xz} = \varphi_x + \frac{\partial w}{\partial x} - \frac{u_0}{R}$$

(4e)

$$k_x = \frac{\partial \varphi_x}{\partial x} \tag{4f}$$

$$k_\theta = \frac{1}{R}\frac{\partial \varphi_\theta}{\partial \theta} \tag{4g}$$

$$k_{x\theta} = \frac{\partial \varphi_\theta}{\partial x} + \frac{1}{R}\frac{\partial \varphi_x}{\partial \theta} \tag{4h}$$

in which R is the middle-surface radius of shells.

The temperature difference between the FGM shell and curing area is given in the following equation:

$$\Delta T = T_0(x,\theta,t) + \frac{z}{h^*}T_1(x,\theta,t) \tag{5}$$

in which T_0 and T_1 are temperature parameters in functions of x, θ and t, h^* is the total thickness of shells.

The dynamic equations of motion for a circular cylindrical shell are introduced by Jafari, Khalili, and Azarafza (2005). The constitutive relations including thermal loads effect are introduced by Lee et al. (2006). The dynamic equilibrium differential equations of FGM circular cylindrical shells in terms of the following stiffness (A_{ij}, B_{ij}, D_{ij}, $A_{i^*j^*}$) with derivatives of displacements and shear rotations, the pulsating axial load and moment (N_a, M_a), external thermal loads expressions ($f_1, \ldots, f_5$), and external pressure load (q) and inertia products of (ρ, H, I) with accelerations terms can be obtained.

$$f_1 = \frac{\partial \bar{N}_x}{\partial x} + \frac{1}{R}\frac{\partial \bar{N}_{x\theta}}{\partial \theta}, f_2 = \frac{\partial \bar{N}_{x\theta}}{\partial x} + \frac{1}{R}\frac{\partial \bar{N}_\theta}{\partial \theta}, f_3 = q - \frac{\bar{N}_\theta}{R}, f_4 = \frac{\partial \bar{M}_x}{\partial x} + \frac{1}{R}\frac{\partial \bar{M}_{x\theta}}{\partial \theta}, f_5 = \frac{\partial \bar{M}_{x\theta}}{\partial x} + \frac{1}{R}\frac{\partial \bar{M}_\theta}{\partial \theta}$$

$$(\bar{N}_x, \bar{M}_x) = \int_{-\frac{h^*}{2}}^{\frac{h^*}{2}} \left(\bar{Q}_{11}\alpha_x + \bar{Q}_{12}\alpha_\theta + \bar{Q}_{16}\alpha_{x\theta}\right) \Delta T(1,z)dz$$

$$(\bar{N}_\theta, \bar{M}_\theta) = \int_{\frac{-h^*}{2}}^{\frac{h^*}{2}} \left(\bar{Q}_{12}\alpha_x + \bar{Q}_{22}\alpha_\theta + \bar{Q}_{26}\alpha_{x\theta}\right) \Delta T(1,z)dz$$

$$(\bar{N}_{x\theta}, \bar{M}_{x\theta}) = \int_{\frac{-h^*}{2}}^{\frac{h^*}{2}} \left(\bar{Q}_{16}\alpha_x + \bar{Q}_{26}\alpha_\theta + \bar{Q}_{66}\alpha_{x\theta}\right) \Delta T(1,z)dz$$

$$(A_{ij}, B_{ij}, D_{ij}) = \int_{\frac{-h^*}{2}}^{\frac{h^*}{2}} \bar{Q}_{ij}(1,z,z^2)dz, \quad (i,j=1,2,6), \quad A_{i^*j^*} = \int_{\frac{-h^*}{2}}^{\frac{h^*}{2}} k_\alpha \bar{Q}_{i^*j^*}dz, \quad (i^*,j^*=4,5)$$

$$(\rho, H, I) = \int_{\frac{-h^*}{2}}^{\frac{h^*}{2}} \rho_0(1,z,z^2)dz, \quad (N_a, M_a) = \int_{-\frac{h^*}{2}}^{\frac{h^*}{2}} (\bar{Q}_{11}\alpha_x + \bar{Q}_{12}\alpha_\theta + \bar{Q}_{16}\alpha_{x\theta})T_0(1,z)dz$$

in which k_α is the shear correction coefficient, ρ_0 is the density of ply. The $\bar{Q}_{ij}$ and $\bar{Q}_{i^*j^*}$ for FGM circular cylindrical shells are used in the simple forms (Sepiani, Rastgoo, Ebrahimi, & Ghorbanpour Arani, 2010).

The simpler stiffness forms of $\bar{Q}_{ij}$ and $\bar{Q}_{i^*j^*}$ are used to calculate the stresses and A_{ij}. For example, the A_{11} integration calculation of FGM shells is given as follows.

$$A_{11} = \frac{h^*}{1 - \left(\frac{v_1 + v_2}{2}\right)^2} \left(\frac{R_n E_1 + E_2}{R_n + 1}\right)$$

(6)

in which E_1 and E_2 are the Young's modulus, v_1 and v_2 are the Poisson's ratios of the FGM constituent layer 1 and 2, respectively.

Usually, the value of shear correction coefficient is not constant in the FGM shells. The modified shear correction factor k_a can be derived based on the energy equivalence principle. Let the total strain energy U due to transverse shears σ_{xz} and $\sigma_{\theta z}$, defined in the following form along the length L of shells by Whitney (1987).

$$U = \frac{R}{2} \int_{-\frac{h^*}{2}}^{\frac{h^*}{2}} \int_{0}^{2\pi} \int_{0}^{L} (\sigma_{xz}\varepsilon_{xz} + \sigma_{\theta z}\varepsilon_{\theta z}) dx d\theta dz$$

(7)

The shear forces Q_θ and Q_x are defined as follows.

$$\left\{\begin{array}{c} Q_\theta \\ Q_x \end{array}\right\} = \left[\begin{array}{cc} A_{44} & 0 \\ 0 & A_{55} \end{array}\right] \left\{\begin{array}{c} \varepsilon_{\theta z} \\ \varepsilon_{xz} \end{array}\right\}$$

(8a)

and also

$$\left\{\begin{array}{c} Q_\theta \\ Q_x \end{array}\right\} = \int_{-\frac{h^*}{2}}^{\frac{h^*}{2}} \left\{\begin{array}{c} \sigma_{\theta z} \\ \sigma_{xz} \end{array}\right\} dz$$

(8b)

Assume for orthotropic, with a constant shear through the thickness of FGM shells, the total strain energy can be rewritten and in terms of the shear correction coefficient k_a.

$$U = \frac{R}{2} \int_0^{2\pi} \int_0^L \left(\frac{Q_x^2}{A_{55}} + \frac{Q_\theta^2}{A_{44}}\right) dx d\theta$$

$$= \frac{R}{2k_a} \int_0^{2\pi} \int_0^L \left(\frac{Q_x^2}{\int_{-\frac{h^*}{2}}^{\frac{h^*}{2}} \bar{Q}_{55} dz} + \frac{Q_\theta^2}{\int_{-\frac{h^*}{2}}^{\frac{h^*}{2}} \bar{Q}_{44} dz}\right) dx d\theta$$

(9a)

The total strain energy U can also be re-expressed in another form in terms of σ_{xz} and $\sigma_{\theta z}$.

$$U = \frac{h^* R}{2} \int_0^{2\pi} \int_0^L \int_{-\frac{h^*}{2}}^{\frac{h^*}{2}} \left(\frac{\sigma_{xz}^2}{\int_{-\frac{h^*}{2}}^{\frac{h^*}{2}} \bar{Q}_{55} dz} + \frac{\sigma_{\theta z}^2}{\int_{-\frac{h^*}{2}}^{\frac{h^*}{2}} \bar{Q}_{44} dz}\right) dz dx d\theta$$

(9b)

By directly integration calculating for σ_{xz} and $\sigma_{\theta z}$ from the static equilibrium equation

$$\sigma_{xz} = -\frac{\partial}{\partial x} \int_{-\frac{h^*}{2}}^{z} \sigma_x dz - \frac{\partial}{R\partial\theta} \int_{-\frac{h^*}{2}}^{z} \sigma_{x\theta} dz$$

(10a)

$$\sigma_{\theta z} = -\frac{\partial}{\partial x} \int_{-\frac{h^*}{2}}^{z} \sigma_{x\theta} dz - \frac{\partial}{R\partial\theta} \int_{-\frac{h^*}{2}}^{z} \sigma_\theta dz$$

(10b)

and for the stresses due to the shear effects of k_x, k_θ, and $k_{x\theta}$.

$$\left\{ \begin{array}{c} \sigma_x \\ \sigma_\theta \\ \sigma_{x\theta} \end{array} \right\} = z \left[\begin{array}{ccc} \bar{Q}_{11} & \bar{Q}_{12} & 0 \\ \bar{Q}_{12} & \bar{Q}_{22} & 0 \\ 0 & 0 & \bar{Q}_{66} \end{array} \right] \left\{ \begin{array}{c} k_x \\ k_\theta \\ k_{x\theta} \end{array} \right\}$$

(10c)

thus

$$\sigma_{xz} = -\frac{FGMZ}{1-v_{fgm}{}^2}\frac{\partial k_x}{\partial x} - \frac{v_{fgm} \cdot FGMZ}{1-v_{fgm}{}^2}\frac{\partial k_\theta}{\partial x} - \frac{1}{2}\frac{FGMZ}{1+v_{fgm}}\frac{\partial k_{x\theta}}{R\partial \theta}$$

(10d)

$$\sigma_{\theta z} = -\frac{1}{2}\frac{FGMZ}{1+v_{fgm}}\frac{\partial k_{x\theta}}{\partial x} - \frac{v_{fgm} \cdot FGMZ}{1-v_{fgm}{}^2}\frac{\partial k_x}{R\partial \theta} - \frac{FGMZ}{1-v_{fgm}{}^2}\frac{\partial k_\theta}{R\partial \theta}$$

(10e)

where v_{fgm} is the Poisson's ratios of the FGM shells.

$$FGMZ = \frac{E_2-E_1}{h^{*R_n}}\left[\frac{\left(z+\frac{h^*}{2}\right)^{R_n+2}}{R_n+2} - \frac{h^*\left(z+\frac{h^*}{2}\right)^{R_n+1}}{2(R_n+1)} \right] + E_1\left(\frac{z^2}{2} - \frac{h^{*2}}{8} \right)$$

The moment resultants M_x, M_θ, and $M_{x\theta}$ are defined as follows.

$$\left\{ \begin{array}{c} M_x \\ M_\theta \\ M_{x\theta} \end{array} \right\} = \int_{-\frac{h^*}{2}}^{\frac{h^*}{2}} \left\{ \begin{array}{c} \sigma_x \\ \sigma_\theta \\ \sigma_{x\theta} \end{array} \right\} z dz$$

(11a)

thus

$$M_x = \frac{FGMZS}{1-v_{fgm}{}^2}k_x + \frac{v_{fgm}FGMZS}{1-v_{fgm}{}^2}k_\theta$$

(11b)

$$M_\theta = \frac{v_{fgm}FGMZS}{1-v_{fgm}{}^2}k_x + \frac{FGMZS}{1-v_{fgm}{}^2}k_\theta$$

(11c)

$$M_{x\theta} = \frac{1}{2}\frac{FGMZS}{1+v_{fgm}}k_{x\theta}$$

(11d)

where

$$FGMZS = \frac{E_2-E_1}{h^{*R_n}}\left[\frac{\left(\frac{h^*}{2}+\frac{h^*}{2}\right)^{R_n+3}}{R_n+3} - \frac{h^*\left(\frac{h^*}{2}+\frac{h^*}{2}\right)^{R_n+2}}{R_n+2} + \frac{h^{*2}\left(\frac{h^*}{2}+\frac{h^*}{2}\right)^{R_n+1}}{4(R_n+1)} \right] + E_1\left[\frac{\left(\frac{h^*}{2}\right)^3}{3} - \frac{\left(-\frac{h^*}{2}\right)^3}{3} \right]$$

Also, the shear forces Q_x and Q_θ can be expressed from the differential equations of equilibrium.

$$Q_x = \frac{\partial M_x}{\partial x} + \frac{\partial M_{x\theta}}{R\partial \theta} = \frac{FGMZS}{1-v_{fgm}{}^2}\frac{\partial k_x}{\partial x} + \frac{v_{fgm}FGMZS}{1-v_{fgm}{}^2}\frac{\partial k_\theta}{\partial x} + \frac{1}{2}\frac{FGMZS}{1+v_{fgm}}\frac{\partial k_{x\theta}}{R\partial \theta}$$

(12a)

$$Q_\theta = \frac{\partial M_{x\theta}}{\partial x} + \frac{\partial M_\theta}{R\partial \theta} = \frac{1}{2}\frac{FGMZS}{1+v_{fgm}}\frac{\partial k_{x\theta}}{\partial x} + \frac{v_{fgm}FGMZS}{1-v_{fgm}{}^2}\frac{\partial k_x}{R\partial \theta} + \frac{FGMZS}{1-v_{fgm}{}^2}\frac{\partial k_\theta}{R\partial \theta}$$

(12b)

By substituting the shear forces Equations 12a and 12b into the total strain energy Equation 9a.

$$U = \frac{R}{2k_\alpha \cdot QBAR} \int_0^{2\pi} \int_0^L \left[\left(\frac{FGMZS}{1-v_{fgm}^2} \frac{\partial k_x}{\partial x} + \frac{v_{fgm}FGMZS}{1-v_{fgm}^2} \frac{\partial k_\theta}{\partial x} + \frac{1}{2}\frac{FGMZS}{1+v_{fgm}} \frac{\partial k_{x\theta}}{R\partial \theta} \right)^2 \right.$$
$$\left. + \left(\frac{1}{2}\frac{FGMZS}{1+v_{fgm}} \frac{\partial k_{x\theta}}{\partial x} + \frac{v_{fgm}FGMZS}{1-v_{fgm}^2} \frac{\partial k_x}{R\partial \theta} + \frac{FGMZS}{1-v_{fgm}^2} \frac{\partial k_\theta}{R\partial \theta} \right)^2 \right] dxd\theta \tag{13a}$$

where $QBAR = \dfrac{E_2 - E_1}{2h^{*R_n}(1+v_{fgm})(R_n+1)} \left(\dfrac{h^*}{2} + \dfrac{h^*}{2} \right)^{R_n+1} + \dfrac{E_1}{2(1+v_{fgm})}h^*.$

Also by substituting the transverse shears Equations 10d–10e into the total strain energy Equation 9b.

$$U = \frac{h^*R}{2 \cdot QBAR} \int_0^{2\pi} \int_0^L \int_{-\frac{h^*}{2}}^{\frac{h^*}{2}} \left[\left(\frac{FGMZ}{1-v_{fgm}^2} \frac{\partial k_x}{\partial x} + \frac{v_{fgm}FGMZ}{1-v_{fgm}^2} \frac{\partial k_\theta}{\partial x} + \frac{1}{2}\frac{FGMZ}{1+v_{fgm}} \frac{\partial k_{x\theta}}{R\partial \theta} \right)^2 \right.$$
$$\left. + \left(\frac{1}{2}\frac{FGMZ}{1+v_{fgm}} \frac{\partial k_{x\theta}}{\partial x} + \frac{v_{fgm}FGMZ}{1-v_{fgm}^2} \frac{\partial k_x}{R\partial \theta} + \frac{FGMZ}{1-v_{fgm}^2} \frac{\partial k_\theta}{R\partial \theta} \right)^2 \right] dzdxd\theta \tag{13b}$$

Assume that $\dfrac{\partial k_x}{\partial x} = \dfrac{\partial k_\theta}{\partial x} = \dfrac{\partial k_{x\theta}}{\partial x} = \dfrac{\partial k_x}{R\partial \theta} = \dfrac{\partial k_\theta}{R\partial \theta} = \dfrac{\partial k_{x\theta}}{R\partial \theta}$, the total strain energy Equation 13a can be derived as follows.

$$U = \frac{R}{2k_\alpha \cdot QBAR} FGMZSV \int_0^{2\pi} \int_0^L \left(\frac{\partial k_x}{\partial x} \right)^2 dxd\theta \tag{13c}$$

where

$$FGMZSV = \frac{(FGMZS)^2}{\left(1-v_{fgm}^2\right)^2} + \frac{v_{fgm}^2(FGMZS)^2}{\left(1-v_{fgm}^2\right)^2} + \frac{(FGMZS)^2}{4(1+v_{fgm})^2} + 2\frac{v_{fgm}}{1-v_{fgm}^2}\frac{(FGMZS)^2}{1-v_{fgm}^2}$$
$$+ \frac{(FGMZS)^2}{(1-v_{fgm}^2)(1+v_{fgm})} + \frac{v_{fgm}}{1-v_{fgm}^2}\frac{(FGMZS)^2}{1+v_{fgm}} + \frac{(FGMZS)^2}{4(1+v_{fgm})^2} + \frac{v_{fgm}^2(FGMZS)^2}{(1-v_{fgm}^2)^2}$$
$$+ \frac{(FGMZS)^2}{(1-v_{fgm}^2)^2} + \frac{v_{fgm}}{1-v_{fgm}^2}\frac{(FGMZS)^2}{1+v_{fgm}} + \frac{(FGMZS)^2}{(1+v_{fgm})(1-v_{fgm}^2)} + 2\frac{v_{fgm}}{1-v_{fgm}^2}\frac{(FGMZS)^2}{1-v_{fgm}^2}$$

Also, the total strain energy Equation 13b can be derived as follows.

$$U = \frac{h^*R}{2 \cdot QBAR} FGMZIV \int_0^{2\pi} \int_0^L \left(\frac{\partial k_x}{\partial x} \right)^2 dxd\theta \tag{13d}$$

where

$$FGMZIV = \frac{FGMZI}{\left(1-v_{fgm}^2\right)^2} + \frac{v_{fgm}^2 FGMZI}{\left(1-v_{fgm}^2\right)^2} + \frac{FGMZI}{4(1+v_{fgm})^2} + 2\frac{v_{fgm}}{1-v_{fgm}^2}\frac{FGMZI}{1-v_{fgm}^2} + \frac{FGMZI}{(1-v_{fgm}^2)(1+v_{fgm})}$$
$$+ \frac{v_{fgm}}{1-v_{fgm}^2}\frac{FGMZI}{1+v_{fgm}} + \frac{FGMZI}{4(1+v_{fgm})^2} + \frac{v_{fgm}^2 FGMZI}{(1-v_{fgm}^2)^2} + \frac{FGMZI}{(1-v_{fgm}^2)^2} + \frac{v_{fgm}}{1-v_{fgm}^2}\frac{FGMZI}{1+v_{fgm}}$$
$$+ \frac{FGMZI}{(1+v_{fgm})(1-v_{fgm}^2)} + \frac{FGMAGZI}{1-v_{fgm}^2} + 2\frac{v_{fgm}}{1-v_{fgm}^2}\frac{FGMZI}{1-v_{fgm}^2}$$

in which

$$
\begin{aligned}
FGMZI = \left(\frac{E_2-E_1}{h^{*R_n}}\right)^2 & \left[\frac{\left(\frac{h^*}{2}+\frac{h^*}{2}\right)^{2R_n+5}}{(R_n+2)^2(2R_n+5)} - h^* \frac{\left(\frac{h^*}{2}+\frac{h^*}{2}\right)^{2R_n+4}}{(R_n+1)(R_n+2)(2R_n+4)} \right. \\
& \left. + h^{*2} \frac{\left(\frac{h^*}{2}+\frac{h^*}{2}\right)^{2R_n+3}}{4(R_n+1)^2(2R_n+3)} \right] + \frac{2(E_2-E_1)}{h^{R_n}} \left\{ \frac{E_1}{2(R_n+2)} \left[\frac{\left(\frac{h}{2}+\frac{h}{2}\right)^{R_n+5}}{R_n+5} - h \frac{\left(\frac{h}{2}+\frac{h}{2}\right)^{R_n+4}}{R_n+4} \right. \right. \\
& \left. + \frac{h^2}{4} \frac{\left(\frac{h}{2}+\frac{h}{2}\right)^{R_n+3}}{R_n+3} \right] - \frac{h^*E_1}{4(R_n+1)} \left[\frac{\left(\frac{h^*}{2}+\frac{h^*}{2}\right)^{R_n+4}}{R_n+4} - h^* \frac{\left(\frac{h^*}{2}+\frac{h^*}{2}\right)^{R_n+3}}{R_n+3} + \frac{h^{*2}}{4} \frac{\left(\frac{h^*}{2}+\frac{h^*}{2}\right)^{R_n+2}}{R_n+2} \right] \right\} \\
& + \frac{E_1^2 h^{*5}}{320} + 2 \left(-\frac{E_1 h^{*2}}{8} \right) \left\{ \frac{E_2-E_1}{h^{*R_n}} \left[\frac{\left(\frac{h^*}{2}+\frac{h^*}{2}\right)^{R_n+3}}{(R_n+2)(R_n+3)} - h^* \frac{\left(\frac{h^*}{2}+\frac{h^*}{2}\right)^{R_n+2}}{2(R_n+1)(R_n+2)} \right] + \frac{E_1 h^{*3}}{24} \right\} \\
& + \left(-\frac{E_1 h^{*2}}{8} \right)^2 h^*
\end{aligned}
$$

By equaling the total strain energy Equations 13c–13d, the shear correction coefficient can be obtained as follows.

$$
k_\alpha = \frac{1}{h^*} \frac{FGMZSV}{FGMZIV} \tag{14}
$$

In 1990, the GDQ method is presented by Shu and Richards and can be re-stated in the following statements: the derivative of a smooth function at a discrete point in a domain can be discrete by using an approximated weighting linear sum of the function values at all the discrete points in the direction (Bert, Jang, & Striz, 1989; Hong, 2007, 2014; Shu & Du, 1997). The GDQ method is used to approximate the derivative of function in the dynamic equilibrium differential equations for four sides simply supported, non-symmetric FGM shells, thus the dynamic GDQ discrete equations in matrix notation can be obtained.

3. Some numerical results and discussion

The following coordinates x_i and θ_j for the grid points of FGM thick circular cylindrical shells are used to study the GDQ results of varied shear correction coefficient calculations with shell layers in the stacking sequence (0°/0°), under four sides simply supported boundary condition, no pulsating axial load and moment ($N_a=M_a=0$), and no external pressure load ($q=0$).

$$
x_i = 0.5 \left[1 - \cos\left(\frac{i-1}{N-1}\pi\right) \right] L, \quad i=1,2,\dots,N \tag{15a}
$$

$$
\theta_j = 0.5 \left[1 - \cos\left(\frac{j-1}{M-1}\pi\right) \right] 2\pi, \quad j=1,2,\dots,M \tag{15b}
$$

The time sinusoidal displacement and temperature of thermal vibrations are used as follows.

$$
u = [u_0(x,\theta) + z\varphi_x(x,\theta)] \sin(\omega_{mn}t) \tag{16a}
$$

$$
v = [v_0(x,\theta) + z\varphi_\theta(x,\theta)] \sin(\omega_{mn}t) \tag{16b}
$$

$$
w = w(x,\theta) \sin(\omega_{mn}t) \tag{16c}
$$

$$
\varphi_x = \varphi_x(x,\theta) \sin(\omega_{mn}t) \tag{16d}
$$

$$\varphi_{\theta} = \varphi_{\theta}(x, \theta)\sin(\omega_{mn}t) \tag{16e}$$

$$\Delta T = \left[T_0(x, \theta) + \frac{z}{h^*}T_1(x, \theta)\right]\sin(\gamma t) \tag{17a}$$

and with the simple vibration of temperature parameter

$$T_0(x, \theta) = 0, \quad (N_a = M_a = 0) \tag{17b}$$

$$T_1(x, \theta) = \bar{T}_1\sin(\pi x/L)\sin(\pi \theta) \tag{17c}$$

in which ω_{mn} is the natural frequency of the shells, γ is the frequency of applied heat flux, $\bar{T}_1$ is the amplitude of temperature.

The FGM layer 1 is SUS304 (stainless steel), the FGM layer 2 is Si_3N_4 (silicon nitride) used for the numerical GDQ computations. Firstly, the dynamic convergence study of center deflection amplitude $w(L/2, 2\pi/2)$ (unit mm) in circular cylindrical FGM shells are obtained in Table 1 by considering the varied effects of shear correction coefficient and with $h^*=1.2$ mm, $h_1=h_2=.6$ mm, $m=n=1$, $R_n=1$, $k_a=.101452$, $T=653°K$, $\bar{T}_1=1°K$, $t=.1$ s. The accuracy is 5.5E-06 for the center deflection amplitude of $L/R=1.0$ and $L/h^*=5$. The 19×19 grid point can be treated in the good convergence result and used in the following GDQ computations of time responses for deflection and stress of circular cylindrical FGM shells. In the circular cylindrical FGM shell ($B_{ij}\neq0$), varied values of k_a are usually functions of R_n and T. For h^* from .12 to 12 mm, $h_1=h_2$, calculated values of k_a under $T=653°K$ are shown in Table 2a, under

Table 1. Dynamic convergence of circular cylindrical FGM shells

L/h^*	GDQ method	Deflection $w(L/2, 2\pi/2)$ at $t = .1$ s		
	$N \times M$	L/R	$L/R = 1.0$	$L/R = 2.0$
100	15×15	.156207E-04	.173512E-04	.301047E-04
	17×17	.155691E-04	.161893E-04	.262582E-04
	19×19	.155689E-04	.161896E-04	.262594E-04
14	15×15	.149683E-01	.402883E-02	.706476E-03
	17×17	.150152E-01	.418445E-02	.727735E-03
	19×19	.150130E-01	.418446E-02	.727721E-03
10	15×15	.106693E-01	.139533E-02	.119783E-02
	17×17	.106911E-01	.143022E-02	.121843E-02
	19×19	.106915E-01	.143028E-02	.121842E-02
8	15×15	.810748E-02	.197729E-02	.160317E-02
	17×17	.811873E-02	.201220E-02	.162178E-02
	19×19	.811960E-02	.201226E-02	.162175E-02
5	15×15	.437213E-02	.357983E-02	2.55451
	17×17	.436597E-02	.360820E-02	2.56697
	19×19	.436321E-02	.360822E-02	2.56694

Table 2a. Varied shear correction coefficient k_a vs. R_n under $T=653°K$

h^* (mm)	k_a						
	$R_n=.1$	$R_n=.2$	$R_n=.5$	$R_n=1$	$R_n=2$	$R_n=5$	$R_n=10$
.12	.811789E-01	.103232	.192780	.332648	.354855	.288468	.255475
1.2	.661600E-01	.693373E-01	.805978E-01	.101452	.138729	.190785	.201838
2.4	.621475E-01	.612361E-01	.595911E-01	.572452E-01	.487517E-01	.165935E-01	.935386E-03
12	.536615E-01	.456055E-01	.283776E-01	.127610E-01	.224098E-02	.563388E-05	.961355E-10

Table 2b. Varied shear correction coefficient k_α vs. R_n under $T=100°K$							
h^* (mm)	k_α						
	$R_n=.1$	$R_n=.2$	$R_n=.5$	$R_n=1$	$R_n=2$	$R_n=5$	$R_n=10$
.12	.107357	.135316	.245718	.403544	.411215	.324596	.283004
1.2	.877684E-01	.914646E-01	.104760	.129271	.171357	.223472	.229368
2.4	.825144E-01	.809008E-01	.777436E-01	.736513E-01	.619167E-01	.209678E-01	.117697E-02
12	.713736E-01	.604279E-01	.372274E-01	.165804E-01	.288825E-02	.717291E-05	.121025E-09

Table 2c. Varied shear correction coefficient k_α vs. R_n under $T=1000°K$							
h^* (mm)	k_α						
	$R_n=.1$	$R_n=.2$	$R_n=.5$	$R_n=1$	$R_n=2$	$R_n=5$	$R_n=10$
.12	.738683E-01	.942588E-01	.190746	.406255	.429641	.289613	.234786
1.2	.591680E-01	.606934E-01	.676897E-01	.831104E-01	.113958	.160412	.167052
2.4	.553258E-01	.530777E-01	.486694E-01	.441184E-01	.353190E-01	.108169E-01	.564123E-03
12	.473142E-01	.387971E-01	.222657E-01	.928170E-02	.153014E-02	.359565E-05	.579053E-10

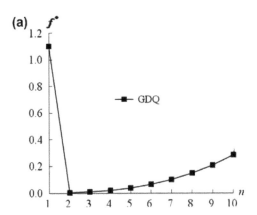

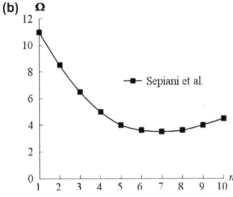

Figure 2. Frequency parameter f^* vs. n and compared Ω vs. n. (a) GDQ result of f^* vs. n; (b) Compared result of Ω vs. n.

$T=100°K$ are shown in Table 2b and under $T=1,000°K$ are shown in Table 2c, respectively, used for the GDQ and shear calculations. For $h^*=12$ mm, values of k_α (from .661600E-01 to .201838 under $T=653°K$, from .877684E-01 to .229368 under $T=100°K$, from .591680E-01 to .167052 under $T=1000°K$) are increasing with R_n (from .1 to 10).

Secondly, the amplitude of center deflection $w(L/2, 2\pi/2)$ (unit mm) for the FGM circular cylindrical shell is calculated. Figure 2a shows the results of frequency parameter f^* vs. n studied by using the

(a) $w(L/2, 2\pi/2)$

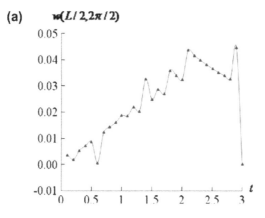

(b) $w(L/2, 2\pi/2)$

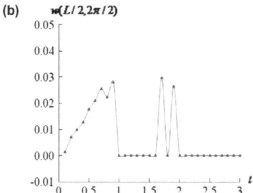

(c) $w(L/2, 2\pi/2)$

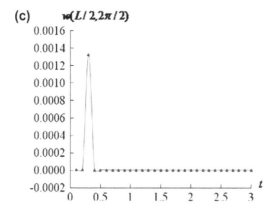

Figure 3. w(L/2, 2π/2) vs. t for L/h* = 5, 10 and 100.

GDQ method for circular cylindrical FGM shells with $h^*=1.2$ mm, $h_1=h_2=.6$ mm, $L/h^*=5$, $L/R=2$, $R_n=1$, the calculated value of shear correction $k_a=.101452$, $T=653°K$, $\bar{T}_1=1°K$, in which $f^* = 4\pi\omega_{mn}R\sqrt{I/A_{11}}$. The values of f deeply decreasing from 1.10157 to .002971 with n increasing from 1 to 2, then f gently increasing from .002971 to .285746 with n increasing from 2 to 10. Figure 2b shows the re-plotted, compared frequency parameter Ω vs. n by Sepiani et al. (2010) with the constant correction factor value $5/6=.833333$ and equation $\Omega=4\pi\omega_{mn}R\sqrt{I_1/A_{11eff}}$, in which $A_{11eff}=\int_{-\frac{h^*}{2}}^{\frac{h^*}{2}} Q_{11}dz$, $I_1=\int_{-\frac{h^*}{2}}^{\frac{h^*}{2}} z\rho(z)dz$, $Q_{11}=\frac{E_{eff}}{1-\nu_{eff}^2}$ with $\rho(z)$, E_{eff}, and ν_{eff} are effective mass density, effective elastic modu-lus, and effective Poisson's ratio of FGM, respectively, for simply supported silicon nitride–nickel FGM cylindrical shell with FSDT under axial extensional loading and no external pressure (q=0). The com-pared results of Figure 2 can be treated in similarly tendency between the two curves of different frequency parameters vs. n.

Figure 3 shows the response values of center deflection amplitude $w(L/2, 2\pi/2)$ (unit mm) vs. time t in FGM circular cylindrical shell for thick $L/h^{\cdot}=5$, 10 and thin $L/h^{\cdot}=100$, respectively, $L/R=1$, $h^{\cdot}=1.2$ mm, $h_1=h_2=.6$ mm, $R_n=1$, $k_a=.101452$, $T=653°$K, $\bar{T}_1=1°$K, $t=.1$–3.0 s, time step is $.1$ s. The maximum value of center deflection amplitude is $.0446$ mm occurs at $t=2.9$ s for thick $L/h^{\cdot}=5$. Figure 4 shows the time responses of the dominated stresses σ_x (unit GP$_a$) at center position of outer surface $z=.5h^{\cdot}$ as the analyses of deflection case in Figure 3 for thick $L/h^{\cdot}=5$, 10 and thin $L/h^{\cdot}=100$, respectively. The maximum absolute value of stresses σ_x is 9.12E-06 GP$_a$ occurs in the periods $t=.2$–3.0 s for thin $L/h^{\cdot}=100$. The absolute values of stresses σ_x are increasing with the values of $L/h^{\cdot}$ are increasing (absolute $\sigma_x=8.4$E-06 GP$_a$ at $L/h^{\cdot}=5$, absolute $\sigma_x=9.12$E-06 GP$_a$ at $L/h^{\cdot}=100$).

Figure 5 shows the center deflection amplitude $w(L/2, 2\pi/2)$ (unit mm) vs. T for all different values R_n (from $.1$ to 10) of FGM circular cylindrical shell calculated and varied values of k_a, for $L/h^{\cdot}=5$, $L/R=1.0$, $h^{\cdot}=1.2$ mm, $h_1=h_2=.6$ mm, $\bar{T}_1=1°$K, at $t=3$ s. The maximum value of center deflection amplitude is 2.23467 mm occurs at $T=1000°$K for $R_n=.1$. The center deflection amplitude values are all

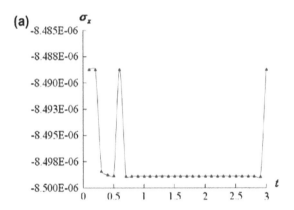

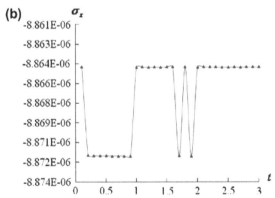

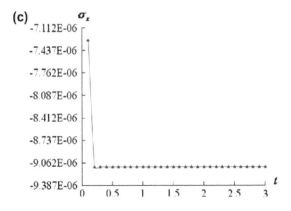

Figure 4. σ_x vs. t for $L/h^{\cdot}=5, 10$ and 100.

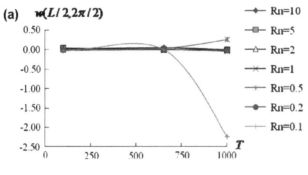

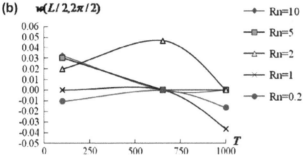

Figure 5. $w(L/2, 2\pi/2)$ vs. T.

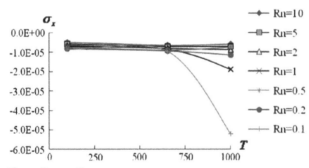

Figure 6. σ_x vs. T.

increasing vs. T for $R_n = .1$ and .5, they cannot withstand higher temperature ($T = 1,000°K$) of environment. The center deflection amplitude value is increasing (from .0197003 to .0467759 mm) and then decreasing (from .0467759 to .000118722 mm) vs. T for $R_n = 2$, it can withstand higher temperature ($T = 1000°K$) of environment. The center deflection amplitude values are all decreasing vs. T for $R_n = 5$ and 10, they also can withstand higher temperature ($T = 1000°K$) of environment. The variant of center deflection amplitude $w(L/2, 2\pi/2)$ for $R_n = .2, 1, 2, 5,$ and 10 are smaller than for $R_n = .1$ and .5. Figure 6 shows the dominated stresses σ_x (unit GP_a) at center position of outer surface $z = .5h^*$ vs. T for all different values R_n of FGM circular cylindrical shell as the analyses of deflection case in Figure 5. The absolute values of dominated stresses σ_x are all increasing vs. T for $R_n = .1$ and 1. Each keeping almost constant stresses values vs. T for others values of R_n.

4. Conclusions

The GDQ solutions are calculated and investigated for the deflections and stresses in the thermal vibration of FGM thick circular cylindrical shells by considering the varied effects of shear correction coefficient. GDQ results show: (1) varied values of k_α are usually functions of R_n and T; (2) The maximum value of center deflection amplitude is .0446 mm occurs at $t = 2.9$ s for thick $L/h^* = 5$ at $T = 653°K$; (3) The maximum value of center deflection amplitude is 2.23467 mm occurs at $T = 1000°K$ for $L/h^* = 5, R_n = .1$; and (4) The center deflection amplitude values are all decreasing vs. T for $R_n = 5$ and

10, they also can withstand for higher temperature ($T = 1,000°K$) of environment. The significance of the results show that for the non-symmetrical and thick thickness of FGM shells, the implantation of calculated and varied shear correction is needed to be considered.

Funding
The authors received no direct funding for this research.

Author details
C.C. Hong[1]
E-mail: cchong@mail.hust.edu.tw
[1] Department of Mechanical Engineering, Hsiuping University of Science and Technology, Taichung 412-80, Taiwan, ROC.

References
Bert, C. W., Jang, S. K., & Striz, A. G. (1989). Nonlinear bending analysis of orthotropic rectangular plates by the method of differential quadrature. *Computational Mechanics, 5*, 217–226. http://dx.doi.org/10.1007/BF01046487

Chi, S. H., & Chung, Y. L. (2006). Mechanical behavior of functionally graded material plates under transverse load. Part I: Analysis. *International Journal of Solids and Structures, 43*, 3657–3674. http://dx.doi.org/10.1016/j.ijsolstr.2005.04.011

Chróścielewski, J., Pietraszkiewicz, W., & Witkowski, W. (2010). On shear correction factors in the non-linear theory of elastic shells. *International Journal of Solids and Structures, 47*, 3537–3545. http://dx.doi.org/10.1016/j.ijsolstr.2010.09.002

Cinefra, M., Belouettar, S., Soave, M., & Carrera, E. (2010). Variable kinematic models applied to free-vibration analysis of functionally graded material shells. *European Journal of Mechanics—A/Solids, 29*, 1078–1087. http://dx.doi.org/10.1016/j.euromechsol.2010.06.001

Du, C., Li, Y., & Jin, X. (2014). Nonlinear forced vibration of functionally graded cylindrical thin shells. *Thin-walled Structures, 78*, 26–36. http://dx.doi.org/10.1016/j.tws.2013.12.010

Ebrahimi, M. J., & Najafizadeh, M. M. (2014). Free vibration analysis of two-dimensional functionally graded cylindrical shells. *Applied Mathematical Modelling, 38*, 308–324. http://dx.doi.org/10.1016/j.apm.2013.06.015

Hong, C. C. (2007). Thermal vibration of magnetostrictive material in laminated plates by the GDQ method. *The Open Mechanics Journal, 1*, 29–37. http://dx.doi.org/10.2174/1874158400701010029

Hong, C. C. (2010). Computational approach of piezoelectric shells by the GDQ method. *Composite Structures, 92*, 811–816. http://dx.doi.org/10.1016/j.compstruct.2009.08.026

Hong, C. C. (2012). Rapid heating induced vibration of magnetostrictive functionally graded material plates. *Transactions of the ASME, Journal of Vibration and Acoustics, 134*, 021019, 1–11.

Hong, C. C. (2013a). Rapid heating induced vibration of circular cylindrical shells with magnetostrictive functionally graded material. *Archives of Civil and Mechanical Engineering.* Retrieved December 2, 2013, from http://www.sciencedirect.com/science/article/pii/S1644966513001441

Hong, C. C. (2013b). Thermal vibration of magnetostrictive functionally graded material shells. *European Journal of Mechanics—A/Solids, 40*, 114–122. http://dx.doi.org/10.1016/j.euromechsol.2013.01.010

Hong, C. C. (2014). Thermal vibration and transient response of magnetostrictive functionally graded material plates. *European Journal of Mechanics—A/Solids, 43*, 78–88. http://dx.doi.org/10.1016/j.euromechsol.2013.09.003

Jafari, A. A., Khalili, S. M. R., & Azarafza, R. (2005). Transient dynamic response of composite circular cylindrical shells under radial impulse load and axial compressive loads. *Thin-walled Structures, 43*, 1763–1786. http://dx.doi.org/10.1016/j.tws.2005.06.009

Lee, S. J., Reddy, J. N., & Rostam-Abadi, F. (2006). Nonlinear finite element analysis of laminated composite shells with actuating layers. *Finite Elements in Analysis and Design, 43*, 1–21. http://dx.doi.org/10.1016/j.finel.2006.04.008

Li, S. R., Fu, X. H., & Batra, R. C. (2010). Free vibration of three-layer circular cylindrical shells with functionally graded middle layer. *Mechanics Research Communications, 37*, 577–580. http://dx.doi.org/10.1016/j.mechrescom.2010.07.006

Malekzadeh, P., Fiouz, A. R., & Sobhrouyan, M. (2012). Three-dimensional free vibration of functionally graded truncated conical shells subjected to thermal environment. *International Journal of Pressure Vessels and Piping, 89*, 210–221. http://dx.doi.org/10.1016/j.ijpvp.2011.11.005

Malekzadeh, P., & Heydarpour, Y. (2012). Free vibration analysis of rotating functionally graded cylindrical shells in thermal environment. *Composite Structures, 94*, 2971–2981. http://dx.doi.org/10.1016/j.compstruct.2012.04.011

Neves, A. M. A., Ferreira, A. J. M., Carrera, E., Cinefra, M., Roque, C. M. C., Jorge, R. M. N., & Soares, C. M. M. (2013). Free vibration analysis of functionally graded shells by a higher-order shear deformation theory and radial basis functions collocation, accounting for through-the-thickness deformations. *European Journal of Mechanics—A/Solids, 37*, 24–34. http://dx.doi.org/10.1016/j.euromechsol.2012.05.005

Qatu, M. S., Sullivan, R. W., & Wang, W. (2010). Recent research advances on the dynamic analysis of composite shells: 2000–2009. *Composite Structures, 93*, 14–31. http://dx.doi.org/10.1016/j.compstruct.2010.05.014

Sepiani, H. A., Rastgoo, A., Ebrahimi, F., & Ghorbanpour Arani, A. (2010). Vibration and buckling analysis of two-layered functionally graded cylindrical shell, considering the effects of transverse shear and rotary inertia. *Materials & Design, 31*, 1063–1069. http://dx.doi.org/10.1016/j.matdes.2009.09.052

Sheng, G. G., & Wang, X. (2013a). Nonlinear vibration control of functionally graded laminated cylindrical shells. *Composites Part B: Engineering, 52*, 1–10. http://dx.doi.org/10.1016/j.compositesb.2013.03.008

Sheng, G. G., & Wang, X. (2013b). An analytical study of the non-linear vibrations of functionally graded cylindrical shells subjected to thermal and axial loads. *Composite Structures, 97*, 261–268. http://dx.doi.org/10.1016/j.compstruct.2012.10.030

Shu, C., & Du, H. (1997). Implementation of clamped and simply supported boundary conditions in the GDQ free vibration analyses of beams and plates. *International Journal of Solids and Structures, 34*, 819–835. http://dx.doi.org/10.1016/S0020-7683(96)00057-1

Strozzi, M., & Pellicano, F. (2013). Nonlinear vibrations of functionally graded cylindrical shells. *Thin-walled Structures, 67*, 63–77. http://dx.doi.org/10.1016/j.tws.2013.01.009

Whitney, J. M. (1987). *Structural analysis of laminated anisotropic plates.* Lancaster, PA: Technomic Publishing.

13

Feasibility of biomass heating system in Middle East Technical University, Northern Cyprus Campus

Samuel Asumadu-Sarkodie[1]* and Phebe Asantewaa Owusu[1]

*Corresponding author: Samuel Asumadu-Sarkodie, Sustainable Environment and Energy Systems, Middle East Technical University, Northern Cyprus Campus, Guzelyurt, Turkey

E-mail: samuel.sarkodie@metu.edu.tr

Reviewing editor: Duc Pham, University of Birmingham, UK

Abstract: Global interest in using biomass feedstock to produce heat and power is increasing. In this study, RETScreen modelling software was used to investigate the feasibility of biomass heating system in Middle East Technical University, Northern Cyprus Campus. Weiss Kessel Multicratboiler system with 2 MW capacity using rice straw biomass as fuel and 10 units of RBI® CB0500 boilers with 144 kW capacity using natural gas as fuel were selected for the proposed biomass heating system. The total cost of the biomass heating project is US\$ 786,390. The project has a pre-tax and after tax internal rate of return (IRR) of 122.70%, simple payback period of 2.54 years, equity payback period of 0.83 year, a net present value of US\$ 3,357,138.29, an annual lifecycle savings of US\$ 262,617.91, a benefit-cost ratio of 21.83, an electricity cost of \$0/kWh and a GHG reduction cost of −204.66 \$/tCO$_2$. The annual GHG emission reduction is 1,283.2 tCO2, which is equivalent to 118 hectares of forest absorbing carbon. The development and adoption of this renewable energy technology will save costs on buying conventional type of heating system and result in a large technical and economic potential for reducing greenhouse gas emissions which will satisfy the sustainable development goals.

Subjects: Bio Energy; Clean Technologies; Energy & Fuels; Environmental; Novel Technologies; Renewable Energy; Renewable Energy; Traditional Industries – Clean & Green Advancements

Keywords: biomass heating; renewable energy; sustainable development; RETScreen; GHG emissions; Northern Cyprus

ABOUT THE AUTHOR

Samuel Asumadu-Sarkodie is a multi-disciplinary researcher who currently studies master's in Sustainable Environment and Energy Systems at Middle East Technical University, Northern Cyprus Campus where he's also a graduate assistant in the Chemistry department. His research interest includes but not limited to: renewable energy, climate change and sustainable development.

PUBLIC INTEREST STATEMENT

The study investigated the feasibility of biomass heating system in Middle East Technical University, Northern Cyprus Campus using RETScreen Clean Energy Project Analysis modelling software designed by Natural Resources Canada. The proposed base load heating system consumes about 1,829 tonnes of rice straw in a Weiss Kessel® Multicratboiler operating at a capacity of 2,000 kW to produce 4,768 MWh of energy for heating purposes at a total peak heating load for all the dormitories assume to be 3,375 kW. The propose biomass heating project costs US\$ 786,390, with an equity payback period of 0.83 year, an annual savings and income of US\$ 330,040 and an annual GHG emission reduction of 1,283.2 tCO$_2$ equivalent to 118 ha of forest absorbing carbon.

1. Introduction

Biomass is ubiquitous and readily available source of energy. The discovery of energy released from wood through fire over one million years BC transformed humanity and civilization (Strezov & Evans, 2014). Accessibility to modern energy services comprises household access to minimum level of electricity; access to safer and more sustainable cooking and heating fuels and stoves; access that enables productive economic activities; and access for public services (International Energy Agency, 2014). The industrial revolution (combustion used to fulfil the basic human needs like: cooking, heating and protection) brought about change of living conditions and technology, and by mid-nineteenth century, technological advancements introduced power stations and the internal combustion engine, requiring a major shift in fuel sources as energy demand increased (Rosillo-Calle, 2012; Strezov & Evans, 2014). The use of biomass decreased and lost its role as the primary source of energy as fossil fuel energy generation gained popularity (Strezov & Evans, 2014). Now, the dominance of fossil fuel-based energy generation in today's increasingly energy-intensive society brings a lot of challenges associated with greenhouse gas emissions (GHGs) (Kabata-Pendias, 2010; Strezov & Evans, 2014): atmospheric pollutants (SO_2, NO_x, particles, traces of metals), water pollution from coal, management of fly ash waste and depletion of fossil fuels. The depletion of fossil fuel and its uneven geographical distribution is drawing fears of energy insecurity, which is a reflection of political instabilities (Cherp et al., 2012; Johansson, 2013; Luft, 2009; Strezov & Evans, 2014). The Intergovernmental Panel on Climate Change (IPCC) report titled: "Climate change 2007: the physical science basis, summary for policy makers", stated that continued GHG emissions from fossil fuels will lead to a temperature increase of between 1.4 and 5.8°C, over the period from 1990 to 2100 (Change, 2007). Falkowski (2000) argues that, understanding the consequences of the aforementioned activities in the coming years is critical for formulating economic, energy, technology, trade and security policies that will affect civilization for generations to come. As a result of this, renewable energy sources like biomass is gaining new global attention, in terms of energy research and development to unearth its advantages to address the growing challenges in energy generation and utilization (Biswas & Kunzru, 2008; Li & Suzuki, 2009; Ng, Lam, Ng, Kamal, & Lim, 2012; Pasini et al., 2011; Tock, Lai, Lee, Tan, & Bhatia, 2010). Global interest in using biomass feedstock to produce heat, power, liquid fuel and chemicals is increasing (Spellman, 2011). Biomass ranks fourth among the sources of energy in the world, which represents about 14% of the world's final energy consumption, higher than coal (12%) and comparable with gas (15%) and electricity (14%) (Demirbas, 2005; Saidur, Abdelaziz, Demirbas, Hossain, & Mekhilef, 2011). Approximately 40% of the world's population relies on traditional bioenergy for their energy needs, accounting for 9% of global energy use and 55% of global wood harvest (Masera, Drigo, Bailis, Ghilardi, & Ruiz-Mercado, 2015). Biomass is the main source of energy in many developing countries, which are mostly non-commercial (Demirbas, 2005; International Energy Agency, 2014; Saidur et al., 2011). Biomass is one of the sources of renewable energy derived from plants through photosynthesis. During photosynthesis, plants combine carbon dioxide and water to form carbohydrates, which constitute the building blocks of biomass (Baskar, Baskar, & Dhillon, 2012). The solar energy that drives the process of photosynthesis is stored in chemical bonds of the carbohydrates and other molecules in the biomass. Biomass is a renewable resource that can be used to generate energy on demand, if it is cultivated and harvested in a manner that allows further growth without depleting nutrients and water resources, with little net additional contribution to global GHGs (Baskar et al., 2012; Hall, Rosillo-Calle, & de Groot, 1992). According to McKendry (2002), "burning new biomass contributes no new carbon dioxide to the atmosphere, as replanting harvested biomass ensures that CO_2 is absorbed and returned for a cycle of new growth". McKendry (2002) defines biomass as any renewable material sourced from a biological origin, which includes anthropogenically modified materials (by-products, products, residues and waste from agriculture, industry and the municipality). The key sources of biomass are: forest residue, whole forest, agricultural residues and crops grown on purpose (Shahrukh et al., 2015).

There are a few literatures on biomass heating systems nonetheless, Li and Wang (2014) investigated the challenges in smart low-temperature district heating development using a holistic approach that measures the reduced system design margin and improve operation of decentralized heat generations. Lund et al. (2014) investigated on the fourth-generation district heating integrating smart thermal grids into future sustainable energy systems. Their study indicated that the fourth-generation district heating involves meeting the challenges of more efficient buildings as well as the operation of smart energy systems. Noussan, Cerino Abdin, Poggio, and Roberto (2014) investigated on the biomass-fired combined heat and power, and heat storage system simulations in existing district heating systems. Torchio (2015) did a comparison of district heating combined heating and power and distributed generation combined heating and power with energy, environmental and economic criteria for Northern Italy. Their study indicated that district heating shows the best values for primary energy savings. Vallios, Tsoutsos, and Papadakis (2009) presented a methodology of the design of biomass district heating systems taking into consideration the optimum design of building structure and urban settlement around the plant. Their study concluded that biomass district heating system model constitutes a favourable and flexible system which responds to all energy requirement in small urban settlements. Wissner (2014) examined the possibility of regulating district heating systems and the difficulties associated with practical implementation using the example of the German district heating market. They concluded that the vast number of district heating generating systems under the European CO_2 trading scheme are economically promising. Ancona, Bianchi, Branchini, and Melino (2014) investigated and analysed the district heating network design using a software based on the Todini–Pilati algorithm generalized by the use of Darcy-Weisbach equation. Their study introduced a district heating network that eliminated combustion systems at final users of thermal energy thereby reducing pollutants and thermal emissions in the study area.

Our study is in line with Stolarski, Krzyżaniak, Warmiński, and Śnieg (2013) who investigated the energy, economic and environmental assessment of heating a family house with biomass. The average consumption of their proposed system ranged from 6.00 to 7.13 t/year at a heat production cost ranging from 713 to 785 €/year. Their study is capable of reducing GHGs from 17.4 to 34.3 t of carbon dioxide equivalent. Nonetheless, their study was limited to a family house which may not be a true representation of the population. Against this backdrop, we present herein, the feasibility of installing a biomass heating system in Middle East Technical University, Northern Cyprus Campus using the RETScreen modelling software by National Resource Canada. An advantage of using RETScreen software is that it facilitates a project evaluation process for decision-making.

2. Materials and method

RETScreen modelling software by National Resource Canada was used to analyse and assess the energy production, energy production cost and savings, GHG reduction, life cycle costs, operation and maintenance cost and financial feasibility of the biomass heating system. Figure 1 shows the biomass heating energy model flowchart. For brevity, not all the equations are outlined since it is already available in RETScreen® International. However, in RETScreen Biomass Heating Project Model, the following mathematical equations are followed (RETScreen® International, xxxx):

$$DD_i = \sum_{k=1}^{N_i}(T_{set} - T_{a,k})$$
(1)

where monthly degree-days is denoted by DD_i, N_i is number of days, set-temperature is denoted as T_{set}, and average daily temperature denoted by $T_{a,k}$.

$$dd_i = \frac{DD_i}{N_i} + dd_{DHW}$$
(2)

where dd_i is the monthly degree-days per day.

The total peak load P_j for the jth cluster of buildings is therefore expressed as:

$$P_j = P_{H,j}A_j$$
(3)

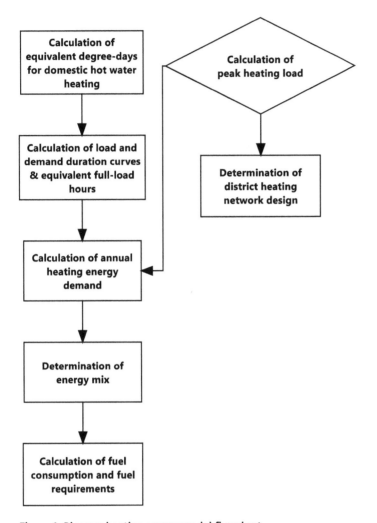

Figure 1. Biomass heating energy model flowchart.

The alternative fuel consumption is calculated as:

$$M_{AFC} = \frac{Q}{\eta_{hs,se} C_f} \tag{4}$$

where M_{AFC} is the alternative fuel consumption, $\eta_{hs,se}$ is the heating system seasonal efficiency, C_f is the calorific value for the selected fuel type and Q is the energy demand of the building or cluster of buildings.

The total heating load carried in a pipe in the main distribution line, P_{pipe}, can be calculated as:

$$P_{pipe} = \rho V C_p \Delta T_{s-r} \tag{5}$$

where ρ is the density of water, V is the volumetric flow of water, C_p its specific heat (78°C, 4,195 J/(kg°C)) and ΔT_{s-r} is the differential temperature between supply and return.

2.1. Proposed biomass heating system

Agricultural biomass namely rice straw bale was selected as a fuel (baseload) for the biomass system. Figure 2 shows the agricultural biomass pathway for heating energy model. Each biomass fuel has a unique characteristic, production amounts per acre, different collection procedure, processing,

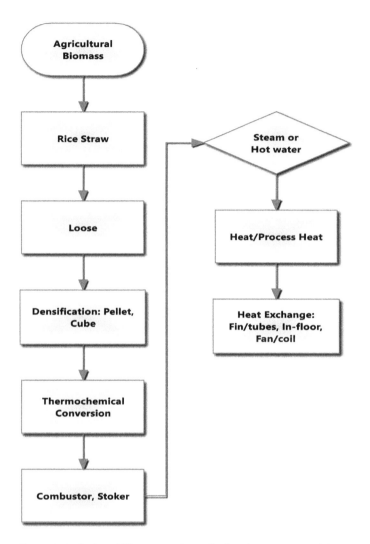

Figure 2. Agricultural biomass pathway for heating energy model.

storage and combustion dynamics (Agricultural Utilization Research Institute, xxxx). The rice straw feedstock for the biomass heating system will require a bale grinder/slicer for processing, a fork lift/ crane as a handling equipment and a barn/shed for storage. The cost of this fuel ranges from $18 (Delivand, Barz, & Gheewala, 2011) to $80 (Agricultural Utilization Research Institute, xxxx) per tonne depending on the location and quality. Rice cultivation is one of the economic activities in Turkey and Northern Cyprus therefore rice straws are enormous and easily available; however, since agricultural biomass sometimes have high levels of alkali in them, fouling issues normally occurs in some biomass boilers (Corp, xxxx). Nonetheless, Weiss Kessel® Multicratboiler biomass system is an exception. Weiss Kessel Multicratboiler system with 2 MW capacity was selected for the proposed biomass heating system. The selection was made because of its multi-purpose function (cyclone, nozzle grate, cyclone burner and push grate furnaces) designed for both dry and wet forms of forest residue, whole forest and agricultural residues. In addition, it has automatic control and regulation system designed for automatic operation, fuel discharging system with infeed and metering systems, the furnace, the boiler and the flue gas cleaning plant until the chimney (WEISS Kessel, 2014).

Natural gas was selected as the fuel for the peak load heating system due to its low cost and its high-reserve discovery in the shore of Northern Cyprus (Caşin, xxxx). The boiler (Model: CB0500) for the system is manufactured by RBI® with a capacity of 144 kW, 98% efficiency, inlet temperature of between 16 and 60°C and a flow rate between 2,322 and 12,609 I/h. The fully modulating firing

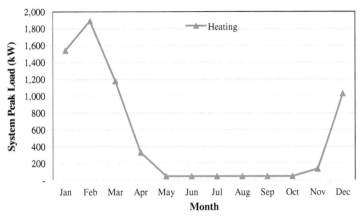

Figure 3. Proposed case system load characteristics graph.

system of RBI® CB0500 continuously varies the energy input to exactly match the heating load without over-firing and wasting fuel providing extremely high part-load efficiencies, this is why this model was selected (RBI, xxxx).

Modelling and simulation of the biomass heating project was performed for three dormitories with two blocks each on Middle East Technical University, Northern Cyprus Campus. The buildings (three dorms with two blocks each) have four floors with a total floor area of about 27,000 m² which accommodates approximately 1,800 students. The total peak heating load for all the dormitories is assumed to be 3,375 kW. Figure 3 shows the proposed case system load characteristics graph. Winter season occurs between November and May in Guzelyurt, Northern Cyprus, corresponding to the increasing heating demand as depicted in Figure 3. The baseload system (primary/main system) can meet the heating demands throughout the year without the peak load system (secondary/reserve system). However, during winter months and coldest days as depicted in Figure 3, the

Table 1. Specifications of the biomass heating system

Item	Parameter
Baseload	*Biomass*
Fuel type	Rice straw
Fuel rate	30 $/t
Biomass system capacity	2,000 kW
Heating delivered	4,768 MWh
Manufacturer	Weiss Kessel®
Model	Multicratboiler
Number of units	1
Seasonal efficiency	80%
Boiler type	Hot water
Fuel required	9.0 GJ/h
Peak load	*Natural gas*
Technology	Boiler
Fuel rate	0.45 $/m³
Capacity	1,440 kW
Heat delivered	196.6 MWh
Manufacturer	RBI®
Model	CB0500
Number of units	10
Seasonal efficiency	80%

baseload system cannot meet the heating demand without the peak load system. Table 1 shows the specifications of the biomass heating system employed in the study.

2.2. Financial analysis

The RETScreen Clean Energy Project Analysis modelling software is capable of performing financial analysis based on financial parameters like; project lifetime, inflation rate, debt interest rate, energy cost, GHG credit, energy cost escalation rate, etc. In Table 2, a summary of financial input parameters used for cost analysis is given. The cost analysis contains a listing of estimated initial and annual cost for biomass heating project. Unless otherwise stated, all the financial input parameters are referred from Minnesota Biomass Heating Feasibility Guide (Agricultural Utilization Research Institute, xxxx).

2.3. Greenhouse gas reduction analysis

According to the IPCC, the annual GHGs can be reduced through technological advancement which requires actions like: adopting energy-efficient technologies and practices, increased fuel switching towards lower carbon fuels, combined heat and power systems, greater reliance on renewable energy sources, etc. (Intergovernmental Panel on Climate Change [IPCC], xxxx). The RETScreen Clean Energy Project Analysis modelling software is capable of performing greenhouse gas reduction analysis based on the energy model of the project. Based on global warming potential of GHG by IPCC 2007, 25 tonnes of CO_2 are equivalent to 1 tonne of CH_4 and 298 tonnes of CO_2 are equivalent to 1 tonne of N_2O (RETScreen, xxxxa). Therefore, the emission factor for CO_2 is 49.4 kg/GJ; CH_4 is 0.0036 kg/GJ, N_2O is 0.0009 kg/GJ and GHG is 0.179 tCO_2/MWh which corresponds to GHG emission of 1,367.1 tCO_2 and a fuel consumption of 7,638 MWh.

The base load heating system of the proposed biomass heating system has the emission factor for CO_2 as 0 kg/GJ; CH_4 as 0.0299 kg/GJ, N_2O as 0.0037 kg/GJ and GHG as 0.007 tCO_2/MWh which corresponds to GHG emission of 39.9 tCO_2 and a fuel consumption of 5,960 MWh.

Table 2. A summary of financial input parameters and assumptions for cost analysis

Item	Value
Initial costs	
Baseload biomass system	US $260/kW
Peak load biomass system	US $50/kW
Feasibility study	US $5,000
Development	US $7,000
Engineering	US $10,000
Miscellaneous/contingency fund	5%
Annual costs	
Parts and labour cost	US $ 10,000
Miscellaneous/contingency fund	5%
Financial parameters	
Debt ratio	75%
Debt interest rate	6%
Debt term	10 years
Fuel cost escalation rate	2%
Discount rate	6%
Inflation rate	2%
Project lifetime	25 years

The peak load heating system of the proposed biomass heating system has the emission factor for CO_2 as 49.4 kg/GJ; CH_4 as 0.0036 kg/GJ, N_2O as 0.0009 kg/GJ and GHG as 0.179 tCO_2/MWh which corresponds to GHG emission of 44 tCO_2 and a fuel consumption of 246 MWh.

3. Results and discussion

In the base case heating system, heating is done using a plant operating on natural gas while the proposed biomass heating system operates on rice straw biomass for the baseload heating system using Weiss Kessel® Multicratboiler and natural gas for the peak load heating system using RBI® CB0500 boiler which serves as a secondary backup during winter months and coldest days. Table 3 shows the summary of the proposed case system. The proposed base load heating system consumes 1,829 tonnes of rice straw in a Weiss Kessel® Multicratboiler operating at a capacity of 2,000 kW to produce 4,768 MWh of energy for heating purposes while the proposed peak load heating system consumes 23,602 m^3 of natural gas in 10 units of RBI® CB0500 boiler operating at a capacity of 144 kW each to produce a total of 197 MWh of energy for heating purposes.

The cost of the project and its savings are critical to its success and investment. The total cost of the biomass heating project for Middle East Technical University, Northern Cyprus campus is US$ 786,390; initial cost is US$ 644,700 which includes feasibility study, development, engineering, the cost of the heating systems and the balance of the system and miscellaneous; the annual costs and debt payment is US$ 141,690 including: operations and maintenance, fuel cost of the proposed case and debt payments for 10 years; and an annual savings and income of US$ 330,040 as a result of avoiding fuel cost in the base case heating system. Table 4 shows the summary of the costs and savings of the proposed project.

Table 5 shows the summary of the financial viability of the proposed project. The first and second row in Table 5 shows the pre-tax and after tax internal rate of return (IRR) of the proposed project. The IRR represents the true interest of the biomass heating system project over its 25 years' lifetime without discount rate assumption (RETScreen, xxxxb). The pre-tax and after tax IRR has the same value of 122.70% as the return on the investment in the biomass heating project.

The third row in Table 5 shows the simple payback period of the proposed project. The simple payback period (Thevenard, Leng, & Martel, 2000), which is the number of years required for the initial cost of the biomass heating project to be paid for out of the savings is 2.54 years.

The fourth row in Table 5 shows the equity payback period of the proposed project. The equity period (Thevenard et al., 2000), which is the time required to recover the equity investment out of pre-tax cash flows reflecting inflation (2%) and debt payments (US$ 65,696) is 0.83 year.

The fifth row in Table 5 shows the net present value (NPV) of the proposed project. The NPV is the sum of all the costs and benefits which is adjusted according to when they occur in the project (Thevenard et al., 2000). The NPV of the project is US$ 3,357,138.29; since the NPV for the biomass heating project is positive, the project is financially attractive at a discount rate of 6%.

The sixth row in Table 5 shows the annual lifecycle savings of the proposed project. The annual lifecycle savings is US$ 262,617.91, which represents the positive savings irrespective of inflation rate, interest rate and taxes.

Table 3. A summary of the proposed case system				
Heating	**Fuel type**	**Fuel consumption**	**Capacity (kW)**	**Energy delivered (MWh)**
Baseload	Rice–straw	1,829 t	2,000	4,768
Peak load	Natural gas	23,602 m^3	1,440	197
		Total	3,440	4,965

Table 4. A summary of costs and savings/income of the proposed project

Initial costs	
Feasibility study	$5,000
Development	$7,000
Engineering	$10,000
Heating system	$592,000
Balance of system & misc	$30,700
Total initial costs	$644,700
Annual costs and debt payments	
O&M	$10,500
Fuel cost–proposed case	$65,494
Debt payments–10 yrs	$65,696
Total annual costs	$141,690
Annual savings and income	
Fuel cost–base case	$330,040
Total annual savings and income	$330,040

The seventh row in Table 5 shows the benefit–cost ratio of the proposed project. The benefit–cost ratio of the biomass heating project is 21.83, which shows a positive value greater than 1 giving a signal that the benefits of the proposed project outweighs its cost.

The eighth row in Table 5 shows the energy production cost of the proposed project. The cost of electricity is $0/kWh.

The ninth and the last row in Table 5 shows the greenhouse gas (GHG) reduction cost of the proposed project. The GHG reduction cost of the project is −204.66 $/tCO$_2$, which means the value of energy saved through greenhouse gas reduction is greater than the capital, operating and maintenance costs (IPCC, xxxx).

Figure 4 shows the cumulative cash flow of the proposed project. It is obvious that the cumulative cash flow is directly proportional to the duration of the biomass heating project, which reassures investors that their profit is secured and in case of unforeseen circumstances, they can still meet their financial obligations at the shortest possible time (Schmidt, 2014).

The proposed biomass heating project has a GHG of 83.9 tCO$_2$ compared to the based case GHG emission of 1,367.1 tCO$_2$. The annual GHG emission reduction is 1,283.2 tCO$_2$, which is equivalent to 118 hectares of forest absorbing carbon.

Table 5. A summary of the financial viability of the proposed project

Financial viability	Unit	Value
Pre-tax IRR–equity	%	122.70
After-tax IRR–equity	%	122.70
Simple payback	yr	2.54
Equity payback	yr	0.83
Net Present Value (NPV)	$	3,357,138.29
Annual life cycle savings	$/yr	262,617.91
Benefit–cost (B–C) ratio		21.83
Energy production cost	$/MWh	0
GHG reduction cost	$/tCO$_2$	−204.66

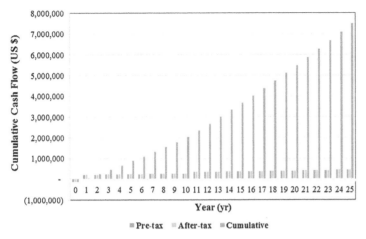

Figure 4. Cumulative cash flow of the proposed project.

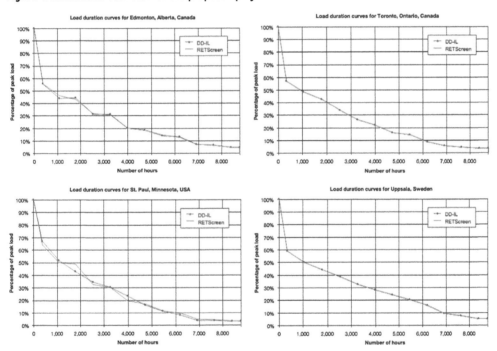

Figure 5. RETScreen model validation against Larsson's model (DD-IL) (RETScreen® International).

3.1. Validation of RETScreen biomass heating system

The load duration curve generated by Project Analysis modelling software designed by Natural Resources Canada has been validated with a computer model by Ingvar Larsson at FVB District Energy Consultants in Sweden. Larsson's model (DD-IL) was developed using extensive and reliable record from a closely monitored District heating systems at Uppsala (Sweden) and St. Paul, Minnesota (USA). The RETScreen Clean Energy Project Analysis model was tested against Larsson's model (DD-IL) with data from four different cities (Edmonton, Alberta (Canada), Toronto, Ontario (Canada), St. Paul, Minnesota (USA) and Uppsala (Sweden)) which showed an accurate and precise output. Figure 5 shows the output of RETScreen model validation against Larsson's model (DD-IL).

4. Conclusions

Global interest in using biomass feedstock to produce heat, power, liquid fuel and chemicals is increasing. In this study, RETScreen Clean Energy Project Analysis modelling software designed by Natural Resources Canada was used to investigate the feasibility of biomass heating system in Middle East Technical University, Northern Cyprus Campus. RETScreen modelling software is capable of analysing and assessing the energy production, energy production cost and savings, GHG reduction, life cycle costs, operation and maintenance cost and financial feasibility of the biomass heating system. Weiss Kessel Multicratboiler system with 2 MW capacity using rice straw biomass as fuel and 10 units of RBI® CB0500 boilers with 144 kW capacity using natural gas as fuel were selected for the proposed biomass heating system. A summary of findings from the study are as follows:

- The total peak heating load for three (two blocks each) dormitories is assume to be 3,375 kW.
- The proposed base load heating system consumes 1,829 tonnes of rice straw in a Weiss Kessel® Multicratboiler operating at a capacity of 2,000 kW to produce 4,768 MWh of energy for heating purposes.
- The proposed peak load heating system consumes 23,602 m³ of natural gas in 10 units of RBI® CB0500 boiler operating at a capacity of 144 kW each to produce a total of 197 MWh of energy for heating purposes.
- The total cost of the biomass heating project for Middle East Technical University, Northern Cyprus campus is US$ 786,390.
- The initial cost of the biomass heating project is US$ 644,700 which includes feasibility study, development, engineering, the cost of the heating systems and the balance of the system and miscellaneous.
- The annual costs and debt payment is US$ 141,690 including: operations and maintenance, fuel cost of the proposed case and debt payments for 10 years.
- The biomass heating system has an annual savings and income of US$ 330,040 as a result of avoiding fuel cost in the base case heating system.
- The project has a pre-tax and after tax IRR of 122.70%, simple payback period of 2.54 years, equity payback period of 0.83 year, a NPV of US$ 3,357,138.29, an annual lifecycle savings of US$ 262,617.91, a benefit–cost ratio of 21.83, a cost of electricity of $0/kWh, and a GHG reduction cost of −204.66 $/tCO$_2$.
- The annual GHG emission reduction is 1,283.2 tCO$_2$, which is equivalent to 118 hectares of forest absorbing carbon.

Biomass heating system is economically feasible in Middle East Technical University, Northern Cyprus Campus. The development and adoption of this renewable energy technology will save costs on buying conventional type of heating system and result in a large technical and economic potential for reducing GHGs which will satisfy the sustainable development goals. Future work should focus on reducing biomass production cost while increasing efficiency.

Nomenclature

Symbols

T_{set}	set temperature
DD	monthly degree-days
N	number of days in a month
Q	annual energy demand
$T_{a,k}$	average daily temperature
Q_H	demand corresponding to space heating
Q_{DHW}	portion of demand corresponding to domestic hot water
D_{DHW}	domestic hot water demand
d	fraction of the annual total demand
C_i	cumulative duration coefficient
D_i	fractions of peak load
F_0	empirical monthly factor

T_{des} design temperature

G_i load duration coefficient

H_i normalize coefficient

E_{flh} equivalent full load hours

P total peak load

A_j total heated area

M_{AFC} alternative fuel consumption

C_f calorific value

Q_{PLHS} peak load heating system

NHV   as-fired calorific value

MCWB   moisture content on a wet basis of biomass (%)

W_{water} weight of water

$W_{drywood}$ weight of dry biomass

HHV   higher heating value (MJ/kg)

P_{pipe} total heating load carried in a pipe

C_p specific heat (78°C, 4,195 J/(kg°C))

ΔT_{s-r} differential temperature between supply and return

Subscript

i month

k' secondary pipe network

k Pipe oversizing factor

j cluster of buildings

hs, se heating system seasonal

bio   biomass

Greek letters

η efficiency

ρ density

Abbreviation

WHR   Waste heat recovery

Acknowledgement
The Authors would like to show their sincere gratitude to the Editor, Duc Pham and the anonymous reviewers for their useful comments. Any errors in the paper are the sole responsibility of the authors.

Funding
The authors received no direct funding for this research.

Author details
Samuel Asumadu-Sarkodie[1]
E-mail: samuel.sarkodie@metu.edu.tr
ORCID ID: http://orcid.org/0000-0001-5035-5983
Phebe Asantewaa Owusu[1]
E-mail: phebe.owusu@metu.edu.tr
ORCID ID: http://orcid.org/0000-0001-7364-1640
[1] Sustainable Environment and Energy Systems, Middle East Technical University, Northern Cyprus Campus, Guzelyurt, Turkey.

References
Agricultural Utilization Research Institute. (xxxx). *Minnesota biomass heating feasibility guide*. Retrieved September 25, 2015, from http://www.auri.org/assets/2012/05/biomass-heating-feasibility-guide.pdf

Ancona, M. A., Bianchi, M., Branchini, L., & Melino, F. (2014). District heating network design and analysis. *Energy Procedia, 45*, 1225–1234. doi:10.1016/j.egypro.2014.01.128

Baskar, C., Baskar, S., & Dhillon, R. S. (2012). *Biomass conversion: The interface of biotechnology, chemistry and materials science*. Springer Science & Business Media. http://dx.doi.org/10.1007/978-3-642-28418-2

Biswas, P., & Kunzru, D. (2008). Oxidative steam reforming of ethanol over Ni/CeO$_2$–ZrO$_2$ catalyst. *Chemical Engineering Journal, 136*, 41–49. http://dx.doi.org/10.1016/j.cej.2007.03.057

Caşin, P. D. M. H. (xxxx). *New energy and peace triangle in the eastern Mediterranean: Israel–Cyprus–Turkey*. Retrieved

September 25, 2015, from http://www.hazar.org/
    blogdetail/blog/new_energy_and_peace_triangle_in_the_
    eastern_mediterranean_israel_cyprus_turkey_629.aspx

Change, I. P. O. C. (2007). Climate change 2007: The physical
    science basis. *Agenda, 6,* 333.

Cherp, A., Adenikinju, A., Goldthau, A., Hughes, L.,
    Jewell, J., Olshanskaya, M., ... Vakulenko, S. (2012). Energy
    and security. In *Global energy assessment: Toward a
    sustainable future* (pp. 325–384).
    http://dx.doi.org/10.1017/CBO9780511793677

Corp, W. F. (xxxx). *Biomass boiler.* Retrieved September 25,
    2015, from http://www.wellonsfei.ca/en/biomass-boiler.
    aspx

Delivand, M. K., Barz, M., & Gheewala, S. H. (2011). Logistics
    cost analysis of rice straw for biomass power generation
    in Thailand. *Energy, 36,* 1435–1441.
    http://dx.doi.org/10.1016/j.energy.2011.01.026

Demirbas, A. (2005). Potential applications of renewable
    energy sources, biomass combustion problems in boiler
    power systems and combustion related environmental
    issues. *Progress in Energy and Combustion Science, 31,*
    171–192. http://dx.doi.org/10.1016/j.pecs.2005.02.002

Falkowski, P. (2000). The global carbon cycle: A test of our
    knowledge of earth as a system. *Science, 290,* 291–296.
    doi:10.1126/science.290.5490.291

Hall, D. O., Rosillo-Calle, F., & de Groot, P. (1992). Biomass
    energy: Lessons from case studies in developing
    countries. *Energy Policy, 20,* 62–73.
    http://dx.doi.org/10.1016/0301-4215(92)90148-U

Intergovernmental Panel on Climate Change. (xxxx). *Working
    group III: Mitigation.* Retrieved September 25, 2015, from
    http://www.ipcc.ch/ipccreports/tar/wg3/index.php?idp=90

International Energy Agency. (2014). World energy outlook
    special report. In I. E. Agency (Ed.), *Africa energy outlook.*

Johansson, B. (2013). A broadened typology on energy and
    security. *Energy, 53,* 199–205.
    http://dx.doi.org/10.1016/j.energy.2013.03.012

Kabata-Pendias, A. (2010). *Trace elements in soils and plants.*
    CRC press. http://dx.doi.org/10.1201/b10158

Li, C., & Suzuki, K. (2009). Tar property, analysis, reforming
    mechanism and model for biomass gasification—An
    overview. *Renewable and Sustainable Energy Reviews, 13,*
    594–604.
    http://dx.doi.org/10.1016/j.rser.2008.01.009

Li, H., & Wang, S. J. (2014). Challenges in smart low-
    temperature district heating development.
    Energy Procedia, 61, 1472–1475. doi:10.1016/j.
    egypro.2014.12.150

Luft, G. (2009). *Energy security challenges for the 21st century:
    A reference handbook.* ABC-CLIO.

Lund, H., Werner, S., Wiltshire, R., Svendsen, S., Thorsen, J. E.,
    Hvelplund, F., & Mathiesen, B. V. (2014). 4th generation
    district heating (4GDH). *Energy, 68,* 1–11. doi:10.1016/j.
    energy.2014.02.089

Schmidt, M. (2014). *Cash flow defined and explained as used in
    finance and accounting.* Retrieved September 26, 2014,
    from https://www.business-case-analysis.com/cash-flow.
    html#cumulative

Masera, O. R., Drigo, R., Bailis, R., Ghilardi, A., & Ruiz-Mercado,
    I. (2015). Environmental burden of traditional bioenergy
    use. *Annual Review of Environment and Resources, 40*(1).
    doi:10.1146/annurev-environ-102014-021318

McKendry, P. (2002). Energy production from biomass (part 1):
    Overview of biomass. *Bioresource Technology, 83,* 37–46.
    http://dx.doi.org/10.1016/S0960-8524(01)00118-3

Ng, W. P. Q., Lam, H. L., Ng, F. Y., Kamal, M., & Lim, J. H. E.
    (2012). Waste-to-wealth: Green potential from palm
    biomass in Malaysia. *Journal of Cleaner Production, 34,*

57–65. http://dx.doi.org/10.1016/j.jclepro.2012.04.004

Noussan, M., Cerino Abdin, G., Poggio, A., & Roberto, R.
    (2014). Biomass-fired CHP and heat storage system
    simulations in existing district heating systems. *Applied
    Thermal Engineering, 71,* 729–735. doi:10.1016/j.
    applthermaleng.2013.11.021

Pasini, T., Piccinini, M., Blosi, M., Bonelli, R., Albonetti, S.,
    Dimitratos, N., ... Kiely, C. J. (2011). Selective oxidation of
    5-hydroxymethyl-2-furfural using supported gold–copper
    nanoparticles. *Green Chemistry, 13,* 2091–2099.
    http://dx.doi.org/10.1039/c1gc15355b

RBI. (xxxx). *Condensing high efficiency domestic water heaters,
    pool heaters and boilers.* Retrieved September 25,
    2015, from http://mesteksa.com/fileuploads/Literature/
    RBI%20Water%20Heaters/Futera%20Fusion/Futera%20
    Fusion%20Catalog%20(FFC-7)11.pdf

RETScreen. (xxxxa). Retrieved September 27, 2015, from http://
    www.retscreen.net

RETScreen. *Retscreen – Introduction – Speaker's notes.* (xxxxb).
    Retrieved September 4, 2015, from http://www.retscreen.
    net/ang/speakers_notes_overview_intro.php

RETScreen® International. (xxxx). Clean energy project
    analysis: RETScreen® engineering & cases textbook.
    Biomass Heating Project Analysis Chapter.

Rosillo-Calle, F. (2012). *The biomass assessment handbook.*
    Earthscan.

Saidur, R., Abdelaziz, E. A., Demirbas, A., Hossain, M. S., &
    Mekhilef, S. (2011). A review on biomass as a fuel for
    boilers. *Renewable and Sustainable Energy Reviews, 15,*
    2262–2289. doi:10.1016/j.rser.2011.02.015

Shahrukh, H., Oyedun, A. O., Kumar, A., Ghiasi, B.,
    Kumar, L., & Sokhansanj, S. (2015). Net energy ratio
    for the production of steam pretreated biomass-
    based pellets. *Biomass and Bioenergy, 80,* 286–297.
    doi:10.1016/j.biombioe.2015.06.006

Spellman, F. (2011). *Forest-based biomass energy: Concepts
    and applications.* CRC Press.
    http://dx.doi.org/10.1201/CRCENERENVI

Stolarski, M. J., Krzyżaniak, M., Warmiński, K., & Śnieg, M.
    (2013). Energy, economic and environmental assessment
    of heating a family house with biomass. *Energy and
    Buildings, 66,* 395–404. doi:10.1016/j.enbuild.2013.07.050

Strezov, V., & Evans, T. J. (2014). *Biomass processing
    technologies.* CRC Press.
    http://dx.doi.org/10.1201/b17093

Thevenard, D., Leng, G., & Martel, S. (2000). The retscreen
    model for assessing potential PV projects. In *Conference
    Record of the Twenty-Eighth IEEE Photovoltaic Specialists
    Conference – 2000* (pp. 1626–1629). doi:10.1109/
    Pvsc.2000.916211

Tock, J. Y., Lai, C. L., Lee, K. T., Tan, K. T., & Bhatia, S. (2010).
    Banana biomass as potential renewable energy resource:
    A Malaysian case study. *Renewable and Sustainable
    Energy Reviews, 14,* 798–805.
    http://dx.doi.org/10.1016/j.rser.2009.10.010

Torchio, M. F. (2015). Comparison of district heating CHP and
    distributed generation CHP with energy, environmental
    and economic criteria for Northern Italy. *Energy
    Conversion and Management, 92,* 114–128. doi:10.1016/j.
    enconman.2014.12.052

Vallios, I., Tsoutsos, T., & Papadakis, G. (2009). Design of
    biomass district heating systems. *Biomass and Bioenergy,
    33,* 659–678. doi:10.1016/j.biombioe.2008.10.009

WEISS Kessel. (2014). Retrieved September 25, 2015,
    from http://www.weiss-kessel.de/cms/show_content.
    php?lang=en&content_id=8

Wissner, M. (2014). Regulation of district-heating systems.
    Utilities Policy, 31, 63–73. doi:10.1016/j.jup.2014.09.001

Permissions

All chapters in this book were first published in CE, by Cogent OA; hereby published with permission under the Creative Commons Attribution License or equivalent. Every chapter published in this book has been scrutinized by our experts. Their significance has been extensively debated. The topics covered herein carry significant findings which will fuel the growth of the discipline. They may even be implemented as practical applications or may be referred to as a beginning point for another development.

The contributors of this book come from diverse backgrounds, making this book a truly international effort. This book will bring forth new frontiers with its revolutionizing research information and detailed analysis of the nascent developments around the world.

We would like to thank all the contributing authors for lending their expertise to make the book truly unique. They have played a crucial role in the development of this book. Without their invaluable contributions this book wouldn't have been possible. They have made vital efforts to compile up to date information on the varied aspects of this subject to make this book a valuable addition to the collection of many professionals and students.

This book was conceptualized with the vision of imparting up-to-date information and advanced data in this field. To ensure the same, a matchless editorial board was set up. Every individual on the board went through rigorous rounds of assessment to prove their worth. After which they invested a large part of their time researching and compiling the most relevant data for our readers.

The editorial board has been involved in producing this book since its inception. They have spent rigorous hours researching and exploring the diverse topics which have resulted in the successful publishing of this book. They have passed on their knowledge of decades through this book. To expedite this challenging task, the publisher supported the team at every step. A small team of assistant editors was also appointed to further simplify the editing procedure and attain best results for the readers.

Apart from the editorial board, the designing team has also invested a significant amount of their time in understanding the subject and creating the most relevant covers. They scrutinized every image to scout for the most suitable representation of the subject and create an appropriate cover for the book.

The publishing team has been an ardent support to the editorial, designing and production team. Their endless efforts to recruit the best for this project, has resulted in the accomplishment of this book. They are a veteran in the field of academics and their pool of knowledge is as vast as their experience in printing. Their expertise and guidance has proved useful at every step. Their uncompromising quality standards have made this book an exceptional effort. Their encouragement from time to time has been an inspiration for everyone.

The publisher and the editorial board hope that this book will prove to be a valuable piece of knowledge for researchers, students, practitioners and scholars across the globe.

List of Contributors

Konstantinos G. Stokos, Socrates I. Vrahliotis and Sokrates Tsangaris
School of Mechanical Engineering, Section of Fluids, Laboratory of Biofluidmechanics & Biomedical Engineering, National Technical University of Athens, Heroon Polytechniou
9, Zografou, 15780 Athens, Greece.

Theodora I. Pappou
FIDES DV-Partner Beratungs-und Vertiebs-GmbH, Munich, Germany

Yongjing Wang
Department of Mechanical Engineering, School of Engineering, University of Birmingham, Birmingham, UK

Lu Wang
Faculty of Engineering and Environment, University of Southampton, Southampton, UK

Mietek A. Brdys
Department of Electronic, Electrical and Systems Engineering, School of Engineering, University of Birmingham, Birmingham, UK
Faculty of Electrical and Control Engineering, Gdansk University of Technology, Gdansk 80-952, Poland

P.U. Suneesh
R&D Centre, Bharathiyar University, Coimbatore, Tamil Nadu, India
Department of Science, MES College of Engineering, Kuttippuram 679573, Kerala, India

John Paul
Department of Science, MES College of Engineering, Kuttippuram 679573, Kerala, India

R. Jayaprakash
Department of Physics, Sri Ramakrishna Mission Vidhyalaya College of Arts and Science, Coimbatore 641020, India

Sanjay Kumar
Department of Physics, BR Ambedkar Bihar University, Muzaffarpur 842001, Bihar, India

David Denkenberger
Civil and Architectural Engineering, Tennessee State University, Knoxville TN, USA

Jean-Luc Rebière
Laboratoire d'Acoustique de l'Université du Maine, (UMR CNRS 6613), Université du Maine, Avenue Olivier Messiaen, 72085 Le Mans Cedex 9, France

B. Ratna Sunil
Department of Mechanical Engineering, Rajiv Gandhi University of Knowledge Technologies (AP-IIIT), Nuzvid 521202, India

Pradeep Kumar Reddy
Department of Mechanical Engineering, Vignana Bharathi Institute of Technology, Hyderabad 501301, India

Priyesh Ray and Sangram Redkar
Department of Engineering, Arizona State University, Mesa, AZ 85212, USA

Tarun Bharath Naine and Manoj Kumar Gundawar
Advanced Centre of Research in High Energy Materials (ACRHEM), University of Hyderabad, Hyderabad 500046, Telangana, India

R. Venkata Rao and Gajanan Waghmare
Department of Mechanical Engineering, S.V. National Institute of Technology, Ichchanath, Surat, Gujarat 395 007, India

R. Selvamani
Department of Mathematics, Karunya University, Coimbatore, Tamil Nadu 641114, India

S.O. Oyedepo
Mechanical Engineering Department, Covenant University, Ota, Nigeria

R.O. Fagbenle
Mechanical Engineering Department, Obafemi Awolowo University, Ile, Ife, Nigeria

S.S. Adefila
Chemical Engineering Department, Covenant University, Ota, Nigeria

Md.Mahbub Alam
Institute for Turbulence-Noise-Vibration Interaction and Control, Shenzhen Graduate School, Harbin Institute of Technology, Shenzhen, China

Jilse Sebastian and Chandrasekharan Muraleedharan
Department of Mechanical Engineering, National Institute of Technology Calicut, Kozhikode 673601, Kerala, India

Arockiasamy Santhiagu
School of Biotechnology, National Institute of Technology Calicut, Kozhikode 673601, Kerala, India

C.C. Hong
Department of Mechanical Engineering, Hsiuping University of Science and Technology, Taichung 412-80, Taiwan, ROC

Samuel Asumadu-Sarkodie and Phebe Asantewaa Owusu
Sustainable Environment and Energy Systems, Middle East Technical University, Northern Cyprus Campus, Guzelyurt, Turkey

Index

Printed in the USA
CPSIA information can be obtained
at www.ICGtesting.com
JSHW051411221024
72173JS00006B/1342